IN THE
WINGS OF PHYSICS

IN THE
WINGS OF PHYSICS

Maurice Jacob

CERN, Geneva
Switzerland

World Scientific
Singapore • New Jersey • London • Hong Kong

Published by

World Scientific Publishing Co. Pte. Ltd.

P O Box 128, Farrer Road, Singapore 9128

USA office: Suite 1B, 1060 Main Street, River Edge, NJ 07661

UK office: 57 Shelton Street, Covent Garden, London WC2H 9HE

Library of Congress Cataloging-in-Publication Data

Jacob, Maurice.
    In the wings of physics / Maurice Jacob.
       p.   cm.
    Includes bibliographical references.
    ISBN 9810221789 (alk. paper)
    1. Physics literature -- Editing.   2. Technical editing.
    3. Physics -- Societies, etc.   I. Title.
    QC5.45.J33   1995
    530'.06--dc20

                                       95-17272
                                        CIP

Printed in Singapore.

INTRODUCTION

"Beauty in things exists in the mind which contemplates them"
David Hume

Physics and Its "Wings"

The theatrical metaphor which serves as a title for this book refers to some of the activities which are going on in the background, or "backstage", of scientific research or, still in other words, in its wings. In the present case, physics is used as a particular example but what is said about physics applies to many other domains. The different activities which are described through this series of articles do not belong to research, as it is usually defined in a restrictive sense, but they are often carried out by people who are also researchers themselves and they take place for the benefit of science. Whereas they may not involve all the excitements of search and discovery and may not bring with them the pleasure and recognition associated with success in research, they contribute much to scientific life. How does the editing of a scientific journal work? How do the so called learned societies operate? The purpose of this book is to take you to such wings of scientific research.

There is certainly a lot more going on in the wings of modern science. Research has become a major endeavour in all the industrialized countries of the world. It employs many people and it is most often organized at the Government level, in the framework of ministries or at least state secretariats. There is definitely a very large amount of work going at those decision and funding levels. The research community is most often involved in such activities, formulating advices, monitoring budgets and implementing new structures. However, it hardly holds any direct executive power. This is a domain where scientific matters and political issues are very much intertwined. There are, however, other backstage activities in which the research community keeps a rather full responsibility, such as those which are focused upon here.

One of them is the domain of scientific publications. Scientist thrive at discovering the secrets of the world. But they are also very eager at quickly making their findings acknowledged by the largest possible interested audience and first of all by their peers or their close colleagues throughout the world.

Publications play a key role in achieving that. It has even become a big business, which is often in the hands of powerful commercial companies. Yet, even in such cases, the research community retains a very large amount of editorial control. We shall get into the inner world of editing and publishing. It is at present at an interesting turning point as electronic publishing is being implemented.

The other types of activity in which we shall get into at some length are those associated with the existence and with the operation of learned societies. These societies are meant to promote science in general terms and they do it with varying successes. They have a long tradition which spans already several centuries but they had to adapt to a changing world. We shall mainly focus on physical societies. They represent a relatively modern form of learned societies since the most senior of them have recently celebrated, or are preparing their hundred anniversary. One important question which often comes up is "What are they useful for?" and I shall try to explain why it is indeed the case. Putting things in a nutshell, let us say that they provide forums for discussion which help at better expressing the views of the scientific communities on many issues of topical importance. These societies are also involved with some scientific publications of their own and, in some cases, these publications are very important. They organize conferences and meetings. All these are activities which are of a fundamental importance to research. Earlier on they were promoting physics in a more general sense and, as part of doing so, awarded prizes. They still do. But we live in a world where funding for research to come has become the key issue and other organizations of a governmental nature are involved in that. Promoting science now implies interacting with those organizations and also addressing the public at large. One should reach much beyond the research community.

Being involved with scientific publications or with physical societies, may be looked at as far less glamorous than contributing to what takes place at the science management level. This may also appear as less exciting than what goes on in the research laboratories, where discoveries take place. Yet they are useful to the life of the scientific community. Whether dealing with scientific publishing or with the various facets of learned societies, one finds interesting challenges and even some fun. Those not familiar with that may therefore find interest in being taken into those parts of the wings of physics where such activities are taking place. The purpose of this book is to make them better known and perhaps better appreciated. They certainly play an important role but their inner working is not very well known, even within the research community.

Scientific publications and learned societies are often specialized. Even though there is a great unity in the scientific approach to the mysteries of the world, research had long to specialize in rather narrow domains, since most individuals are not able to provide new ideas and new results outside of a rather specialized branch of knowledge. It should therefore not be a surprise that this book refers only to one particular science, namely, physics. Within this restricted scope, it is therefore leading the reader only into those wings of research where one deals with publications in physics and with the running of physical societies. Nevertheless, this is already a wide and interesting world and what is said about physics often also applies to many other domains of scientific research with hardly any change.

Presenting these activities is not done in a didactive way. It is rather attempted through a series of papers which were written over a rather large span of time, covering more than twenty years. Some of these papers were published but they appeared in bulletins, journals or books which are not so easily available. Others were just write-ups of talks which were once given. This bring to the subject a strong personal touch but I am happy to be thus able to share through one single book a large amount of a personal experience with editing on the one hand and with the inner sides of physical societies on the other hand.

A Personal Experience

It is impossible to approach physics without passion. Ever since I was in my teens, I have been chafing under the desire to understand natural phenomena and, may be, to add some little things of my own to their understanding. I have been priviledged to spend most of my active life as a research physicist and the more so working in that frontier domain known as particle physics. It had just come of age, with its specific problems, its particular vernacular and its own journals and series of conferences, when I entered the field as a young researcher. This was in the late fifties. It has been just wonderful to live through a time when a new level in the structure of matter, the quark level, was discovered and explored. At present, this is a level where physics manifests a unity and a simplicity which it never achieved so far. This is embodied in the Standard Model of fundamental particles and fundamental interactions. This is a level where the study and understanding of the deep structure of matter brings us to a position where something can be said about the very early universe, when the Big Bang already shaped what was to develop into the many facets of the world we live in.

Physics is made by physicists. As human beings, they organize their work in a human way and this reflects on many aspects of research which have gained in importance as scientific research became more and more a collective effort.

Physics research certainly remains to a large extent an individual endeavour insofar as a particular idea originates in a particular brain. Nevertheless, the genesis of an idea or the recognition of a new effect, often borrows much from contacts and discussions with colleagues. Recognizing the validity and interest of a new idea, or of a new result, or the relevance of a new phenomenon, almost always require the critical assessment and the further contributions of others. Making new results and new ideas quickly known to a wide and interested audience calls for good and efficient means of circulation which usually take the form of journals or conferences. We meet here the role of publications and of physical society.

Developing and exploiting the instrumentation necessary for new forrays often calls for collaborations involving usually many researchers. Selecting the new major instruments to be built demands much preliminary consultation and many critical discussions. Building up the constituency which could only justify the construction of a large facility requires forums for discussions where those to be convinced are not only those likely to be interested in the first place but also colleagues from other domains who may be competing for the same funds, or think they do. We meet here again the role of physical societies.

Physics publications, with their editorial and refereeing problems, and physical societies, with their organizing conferences and forums for discussions, are certainly two important activities in the wings of physics research. Some publications and, in any case, physical societies also try to keep alive the deep unity of physics. This is much needed in a world where extreme specialization is often required for any important advance.

Throughout my research career, I was led to invest much time and effort in such wing activities and this on several occasions and in a variety of ways. Physics research is a passion which is very greedy on time. One is therefore always reluctant to accept responsibilities which take away time which one would better like to devote to one's own research. Getting involved in these backstage activities therefore often results from a lack of ability to say "no" at the proper time. This was certainly the case for me. Nevertheless, having been led to accept such tasks, I did find much challenge and interest in them and it is this experience which I would like to share. I would like to convey the enthusiasm with which I tried to conduct these tasks as editor of journals, on two different occasions, and as president of two physical societies. "Why should you bother to do such things?" was the question which I often heard from those who were not trying to convince me to accept. Well, I nevertheless did accept and, in retrospect, I do not regret it.

Practically, all my career took place in large laboratories, in Europe and in the United States and most of it actually in this wonderful Organization which CERN is. I certainly much benefited from my presence at CERN, as a research physicist, to run my editorial works and also to fulfill my various duties in physical societies. It is my hope that CERN benefited also from it, at least a little.

At the beginning of my research career, I have been very fortunate to have Gian Carlo Wick as my thesis adviser. He was a physicist of impressive insight and ability. He was a man of great culture with a profound humanism which he himself much admired as a young man with Bohr and Sommerfeld. He was very generous of his time with his fellow physicists and indeed part of his impressive work was involved with new and difficult questions of physics which he was able to clarify for the benefit of many. His search for clarity, so that others could understand, was even one of his primary motivation when he derived the famous theorem which goes with his name. Even though I could work with him for a rather short time only, he always remained a prominent example for me as he himself had said about his relation with Fermi. As I was involving myself more and more with such wing activities, I took comfort in the fact that it was a work done to the benefit of the community at large which he would approve, considering it also as valuable and necessary.

It is not my purpose to now reflect in some general ways on these past activities but rather to present them as I lived them, as it is possible through this series of reprints. This is a succession of short papers which I wrote at various times, as the action was going on or as the action had just taken place. This collection of papers certainly stands as a better testimony than a new text which would undoubtedly have mixed too much present views with past experience. I could not have resisted a selection meant at supporting my present pet points of views. All of the papers reprinted in this book are to a large extent independent of one another and they are meant to be read separately and almost in any order. Each one of them was indeed written at a particular time in order to cover a particular point. When written, they were however, each aimed at readers who would be knowledgeable about many of the facts or questions referred to. This may not be the case for present readers. In order to remedy for that, the different papers are now preceeded by a short foreword which brings the relevant issues into perspective. They are otherwise left as they were, with no addition which would make them benefit from the knowledge of what happened afterwards. From their independent writing, and intended independent reading, they show some overlap.

Editing

I was at CERN on a fixed term position in the Theory Division when I was asked by Martin Veltman to succeed him as editor for the theoretical particle physics section of *Physics Letters B*. *Physics Letters B* is a letter journal published by North-Holland. A publishing company operating on its own at that time, it is now part of Elsevier. Today *Physics Letters B* is certainly a well-known journal. It specializes in particle and nuclear physics. Jacques Prentki had been one of its first editors when the journal was launched in the early sixties and he had successfully contributed to put it on the map. In charge of the theoretical particle physics section, he had been succeeded by James Hamilton who had then been succeeded by Martin Veltman. Running a letter journal requires time and energy. The natural duration time for an editor looked like three years. Jacques Prentki, who was then Head of the Theory Division at CERN, where I was working, encouraged me to accept. I had no convincing arguments to oppose to those of Martin Veltman. I did not say "no" and took the job. This was in 1968.

This immediately became what I considered to be an interesting challenge. I wanted to improve the international standing of the journal and in particular, to bring more American authors to submit their papers to it. At that time, the leadership of America in particle physics was still overwhelming and getting the respect of American authors was therefore the real test. By that time, I had already spent three years in America as a research physicist at Brookhaven, at Caltech and then at the Stanford Linear Accelerator Center. I knew many American colleagues well. This helped a lot!

By 1971, I had to look for a job as my contract with CERN was due to terminate. I like working in big laboratories since my work as a theorist has always shown a strong phenomenological slant, with many discussions with experimental colleagues. I always did so with only some graduate teaching assignments in several universities. I therefore left for Fermilab, where I was offered a position. I could not continue with *Physics Letters B* since I had to concentrate first, on my research and second, on my looking for a permanent position as I was already in my late thirties. I was fortunate to be able to convince Raoul Gatto and Euan Squires to succeed me. The input had indeed increased in an important way over the past three years and dealing with all the incoming papers in theoretical particle physics was definitely becoming too much for a single editor. I also had done my three years and felt that this was enough. As I left the editorship of *Physics Letters B*, I wrote a paper attempting to summarize this three years experience with editing. It was accepted in a new journal which eventually did not take off. This journal should have been the new format of *Europhysics News*, the bulletin of the European Physical Society. Here it is, published therefore for the first time, under the title *An editing experiment*. This is Section 1.

By 1970, North-Holland was considering launching a new journal which would carry reviews of recent developments in physics. The idea was to break that way the overspecialization of physics which was too clearly reflected in the current research literature and, in particular, through the articles published in *Physics Letters B* and in *Nuclear Physics B*, which were their journals in the case of particle physics. W. Wimmers, the manager of the Company by then, asked me to become one of the editors of this new journal which eventually was to appear in 1971 under the name of *Physics Reports*.

I had much esteem and admiration for W. Wimmers and was impressed by the way he was dealing with his work. I just could not say "no" to him. I also found it an interesting challenge. Indeed, setting up the format of the new review journal turned out to be a tricky matter. This is described in Section 2. In any case, it was ready to go by the end of 1970. I contributed to giving it a good start in 1971 but had then to quickly slow down as I had to change jobs for the reasons which I previously explained. Editing could no longer be a priority! When I came back to CERN by the end of 1972, with a permanent position this time, I could resume my editorial work fully, dealing by then with most of the particle physics component of the journal, whether on the theoretical side or on the experimental side. I thus worked for *Physics Reports* until 1985. This corresponds to a total of fifteen years, which is really a lot for an editor! One should not reach the point where the work becomes routine. It was indeed time to go, not to impose too much of my own style to a journal. On top of that, in 1985, I had to be at the same time, Head of the Theory Division of CERN, which I had been since 1982, and president of the French Physical Society. I had to stop being also editor of *Physics Reports*. It was also still a great time for hadronic jets and this was not to be missed. This was my pet research theme at that time. I was rapporteur for that field at the 1984 and at the 1990 Rochester Conferences. All this was a lot to face at the same time. As I decided to drop *Physics Reports*, I was happy to be able to convince Roberto Petronzio and Richard Slansky to succeed me as editors for particle physics. For reasons which I have already mentioned, I always felt that the American interest in a European based journal was of cardinal importance for its success. During my fifteen years as an editor the number of European and American papers had remain about equal but, during that time, I had also spent at least one month per year in the United States, mainly at Fermilab, something which greatly helped in my rallying American authors. I always fought, and with some success, for having American physicists among the editors of the North-Holland journals. I was therefore very happy to be able to leave my job to two much esteemed colleagues, one a European and one an American. The current input in particle physics which had no reason to be restrained was in any case such that neither of them could complain to have too little to deal with.

As I left my editorship with *Physics Reports*, I again wrote a paper about that past experience. It is reproduced here under the title *Physics reports: an editorial experiment*. This is Section 2.

It was during those *Physics Reports* years that I came to know about World Scientific. I immediately felt much admiration for the drive and enthusiasm of its thriving managers and I still remember with great pleasure, my first visit to Singapore in 1983 when World Scientific was still at a budding stage. I was happy to follow closely the successful start of WSPC and I am glad to have collaborated with it on quite a few projects by now. Whilst I kept my main editorial allegiance in Amsterdam, I took a particular pride in being at the origin of several joint ventures which took place between World Scientific and North-Holland. I cannot refrain a feeling of pleasure when I see the two logos, one on top of the other, on the spine of a book, and there are indeed several well-cited books like that. These joint projects covered reprint volumes originating from *Physics Reports*, which found this way a paperback version, and some famous *Les Houches* summer school proceedings, which were first published by North-Holland and later reprinted by World Scientific. I indeed felt that the complementarity of the two publishing companies had to be exploited for the benefit of the community. The first publications corresponded to timely and high quality books which, as a result, prices which were often too high for the individual researcher. For titles much in potential demand, publishing a much cheaper paperback reprint version was deemed too small a task for "big" North-Holland when it was naturally bread and butter for the, by then, "small" World Scientific. I could use my positions as an editor of *Physics Reports* on the one hand, and as a long time member of the governing board of the *Les Houches* summer school, on the other hand, to arrange the weddings. At the time, everybody was winning and was happy. Several books thus benefited from their being first published by North-Holland and from their being eventually put at an affordable price to the individual physicists through a World Scientific paperback version.

In 1990, I was invited to give a talk about editing in physics to the Management in Amsterdam, which had much changed since my early time as an editor, in the seventies and early eighties. I felt much honoured by that and wrote up the talk. It is reproduced here under the title *Publishing and Editing: Some reflections on eighteen years as an editor*. This is Section 3.

By that time, I had also gone through another interesting experience of a new kind, in the wings of physics once again, but looking this time at publication from a librarian point of view. It so happened that, during the six years which I spent as Head of the Theory Division of CERN, from 1982 to 1988, the CERN library was put administratively within the Theory Division. This also happened to be just at the time when the head librarian had to leave on retirement. This particular feature had been overlooked by the Management at the proper time. As a result, the library was left without a professional head librarian for a full year before a new one could be recruited. I had therefore to step in, learning as quickly as possible how to pull some of the ropes of the trade. I collected much inspiration from a little book by Umberto Eco called *De Bibliotheca*, which the author of the famous *The Name of the Rose* wrote after a speech which he had been asked to give on the occasion of the opening of a new library in Milan. I bravely gave the same title to a little note I wrote about physics library and which summarizes the guiding lines which I followed during my own library experience. It is also reproduced here under the title *De Bibliotheca*. This is Section 4.

Physics libraries are indeed changing. They no longer take their pride in a collection of books, with spines filling numerous shelves, but in their ability to quickly provide their users with the adequate material, whether in the form of "grey" or bona fide literature and whether it is available on the spot or has to be obtained from somewhere else.

At present, there is much discussion going on about electronic publishing. Some publications are even already operated electronically and this is bound to develop. One could once think that the foreseen limitation in the amount of paper which could be circulated around would put a natural limit to the extension of the scientific literature with, in physics alone, about four hundreds new papers per day! With electronic publishing, there is however, no longer any such end in sight! It remains that the amount of words which any one can read during a particular time is limited. Extracting the signal from the noise could become increasingly difficult. As a result, the larger the amount of circulated material becomes, the more important it is to sort out that material for the benefit of the reader, with the selection and quality control usually offered by refereed journals. Electronic publishing should therefore be considered as a new "hardware", which is able in many ways to substitute paper publishing, with considerable reductions in composition and delivery time and also, may be, in cost. This new publication mean will however, have to incorporate the "software" associated with the editing and refereeing procedure, which actually makes the quality of a particular means of publication. In 1993, I was invited to present my view on the subject at the meeting of the International Federation of Scientific Editors. I wrote up my contribution in a paper which is included here under the title *Editing in physics: Quality control in the electronic age.* This is Section 5, written prior to the development of World Wide Web.

Next to this editorial experience, thus illustrated by papers which cover various facets of my activities in that domain, and which were written from 1971, for the first one, to 1993, for the last one, I would like to talk about my own experience with the inner activities of physical societies, which is another of my ventures in the wings of physics. This is done with another series of papers.

Physical Societies

From 1984 to 1986, I was vice-president and president of the French Physical Society. In this case, once again, it started with my lack of ability at saying "no". When Anatole Abragam asked me whether I would like to be the president of the French Physical Society, I could but say "yes" to him and this, despite the fact that my being Head of the CERN Theory Division was already representing a big toll on my research time. At that time, the tour of duty corresponded to three years. One was vice-president for a year, then president for a year, and then vice-president again for a last year. Being the president of the French Physical Society brought me to the Council of the European Physical Society. I was then elected to its Executive Committee, unable again to say "no" at the proper time. Being on the Executive Committee, I was eventually elected secretary and then president and I served my two successive mandates as president from the spring of 1991 to the spring of 1993. The tour of duty in that position is two years.

This altogether long experience with the inner circles of the French and then of the European physical societies took a fair amount of time and energy but, there again, I met interesting challenges and, at present and in retrospect, I am happy that I did not say "no", even though I sometime cursed myself for having accepted during my presidential years. I thus wish to tell about this experience with the hope to entice others to follow. Indeed, if all physicists are well aware of the needs for physics journals and for editors to run these journals, many of them do not yet perceive well, or even do not perceive at all, the need for physical societies and therefore for presidents of these societies. It is therefore more of a challenge to try to explain why such institutions are useful and actually, as I think, bound to become even more useful. In 1993, I tried to summarize my experience with physical societies in an article which I was invited to write for a book in honour of Victor Weisskopf on the occasion of his eightieth birthday. It is included here under the title of *Physical societies and physics communities*. This is Section 6.

There was a World Scientific venture during my presidency of the French Physical Society. I then took it as a challenge to foster contacts between the French Physical Society and the German Physical Society. I met much good will on both sides and it went very well. We started a joint prize, the Gentner-Kastler prize, which is awarded alternatively, every even year to a French physicist by the German Physical society and every odd year to a German Physicist by the French Physical Society. This was the first prize ever with an amount labelled in Ecus! This is a practice which has since caught up in Europe. We also had a joint meeting on *The Quark Structure of Matter*, which took place in the Rhine valley, with three days in Strasbourg followed by two days in Karlsruhe. The proceedings were published by World Scientific. The first book ever, carrying both the names *Societé Française de Physique* and *Deutsche Physikalische Gesellschaft* on its cover thus turned out to be a World Scientific book!

While writing the paper which now corresponds to Section 6, I could present my views as they were when a full nine years experience with the inner circles of physical societies, an experience extending from 1984 to 1993, was over. In order to stick more closely to the spirit of this book, where papers are reproduced as they were written at a particular time and for a particular purpose, I wish to also include my three main addresses to the European Physical Society, presented in 1991, 1992 and 1993, respectively. They have all been published in *Europhysics News*, which is the bulletin of the European Physical society, but its circulation is still limited. They altogether illustrate the challenges, the hopes and some results, just as they came about. This corresponds to Section 7. This series of addresses is presented in support for the present usefulness, and more importantly, the potential of this Society in the development of European Physics. It is my hope that Sections 6 and 7 can contribute in a positive way to answering the question "What is the use of a physical society?".

Some people often try to separate basic physics from applied physics. This is often the case when funding priorities are assessed with a result which one may easily guess. As physicists, we should insist that no such clear dichotomy exists. There is a full continuity between the most fundamental and the most clearly applied research. It remains that in our present world, when short term economic returns are highly priced, the relation between physics research and industry is an often debated question. This collection ends therefore with, in Section 8, an opening address to the Conference *Physics for Industry, Industry for Physics*, which was held in Cracow, Poland, in 1991. This conference was organized under the joint sponsorship of the Polish and European Physical societies.

CONTENTS

INTRODUCTION   **v**

Part I
Working with Physics Journals

1. An Editing Experiment: Running the Theoretical Particle Physics
   Section of *Physics Letters B*, 1968–1971   **3**
   1.1. Introduction   4
   1.2. Sketch of a Policy   7
   1.3. World Implantation and Working Rate   16
   1.4. A Letter Journal, What Else?   20
2. *Physics Reports:* An Editorial Experiment: Running the Particle
   Physics Section of *Physics Reports*, 1970–1985   **23**
   2.1. Introduction   24
   2.2. The Conception, Birth and Youth of *Physics Reports;*
       A Recollection   25
   2.3. *Physics Reports*, A Few Figures   32
   2.4. The Bulk Order Scheme   37
   2.5. A Few Editorial Comments   40
3. Publishing and Editing: Some Reflections on
   Eighteen Years as an Editor   **45**
   3.1. Introduction   45
   3.2. Publishing in Physics   47
   3.3. Publication Means   49
   3.4. The Publisher   52
   3.5. The Dynamics of Scientific Publishing   53

3.6. The Success of North-Holland in Particle Physics   55

3.7. Physics Libraries   57

4. *De Bibliotheca*: An Essay on Past and Modern Physics
Research Libraries   **61**

    4.1. The Modern Scientific Library   62

    4.2. Where Money Becomes a Problem   64

5. Editing in Physics: Quality Control in the Electronic Age   **69**

    5.1. Introduction   70

    5.2. A Personal Experience   71

    5.3. Publishing and Editing   72

    5.4. The Dynamics of Journals   78

    5.5. Some Needed Steps Toward Electronic Publishing   79

Part II
Working with Physical Societies

6. *Physical Societies and Physics Communities*: A Contribution to
Achievements in Physics, a book in the honour of V. Weisskopf   **85**

    6.1. The World of Physics   86

    6.2. Physical Societies, Roles, and Goals   87

    6.3. Physical Societies in Europe   89

    6.4. Physical Societies and Learned Societies   89

    6.5. The Roles of Physical Societies   92

    6.6. Three Years with the SFP   96

    6.7. The Two Years as President of the EPS   99

7. *Three Addresses to the European Physical Society*   **105**

    a. "Looking Forward". Address of the President — 1991   105

    b. Address to the Council of EPS — 1992   115

    c. Address to the Council of EPS — 1993   125

8. *The EPS, Physics and Industry*   **137**

Part I

Working with Physics Journals

RUNNING THE THEORETICAL PARTICLE PHYSICS SECTION OF *PHYSICS LETTERS B*, 1968–1971

−1971−

From 1968 to 1971, I was an editor of *Physics Letters B*. This is a letter journal which specializes in nuclear and particle physics. They were both highly topical domains in research at that time, as they have remained so, and there was a clear need for a quick and wide circulation of information. *Physics Letters B* had started in 1962. In 1968, there were still four editors, covering respectively theoretical and experimental nuclear physics and theoretical and experimental particle physics. I was in charge of the theoretical particle physics section of the journal. The four editors worked almost independently of one another and there were very few meetings where they could exchange views also with the editors of the *A* section of the journal, which was dealing with all the other domains of physics.

The need for letter journals, which would carry relatively short articles, presenting new and potentially very interesting developments, had manifested itself in the late fifties and the pioneer in that domain had been *Physical Review Letters*, published by the American Physical Society. Papers appearing in a letter journal were quickly attached with some extra prestige, acceptance, meaning the recognition for some urgency at publication.

The difficult role of the editor is to assess quality but also this urgency. This has to be done when faced with a rather large amount of incoming papers. During my time, the input was about one paper per day. It has much increased since! The rejection rate has reluctantly to be high so that the journal would serve the readers as well as the authors. In an ideal situation, it has to remain a must to read for the active researcher in so far as he (she) would be sure to find there, short papers all recognized to be of high topical interest. Including

3

too many papers would please more authors but would result in a bulky journal which could deter the reader. There is, however, certainly no reason not to accept all the original and interesting papers which are submitted and drawing the line is therefore, rather difficult. The only golden rule is that rejecting a good paper is very detrimental to the prestige of the journal and accepting a bad one is even worst. Refereeing and editing have to be quick since urgency at publication is the important issue. Yet the process has to be fair and relatively transparent so that authors can trust the journal.

During my time, an important issue was also to try to attract American papers. There was still an inferiority feeling with respect to America among European researchers and having American authors publishing in the journal was considered as extremely important for its prestige and success. We had therefore to compete with the prestigious *Physical Review Letters*!

When I left my editorial position in 1971, after close to three years on the job, I decided to write a paper about this editing experience which had been new to me. This I did in the style of a research paper. Its purpose was to describe the editorial policy which had been followed and also to give information about the field distribution and the geographical origin of the papers received.

Physics Letters B offers the example of a European journal which could eventually impose itself next to one of the prestigious American. In some circles, one often hears the complaint that publications in physics are too much in "anglo-saxon" hands and that, this is detrimental to researchers in continental Europe or elsewhere in the world. Too many people readily associate publication in English to an American or British origin. The publications of North-Holland offer a beautiful proof that this is not always the case. North-Holland is a commercial publisher but, as an editor, I was entirely free to decide what was to be accepted for publication. Its journals are published in the Netherlands but their editorial boards are very international.

1.1. Introduction

The purpose of this note is to describe the editing procedure followed up with the section of *Physics Letters B* devoted to theoretical elementary particle and high-energy physics[1] during the two-year period extending over 1969 and

---

[1] *Physics Letters* is divided into two Sections, *A* and *B*. The *B* Section is further subdivided into four parts with separate and independent editors. The present discussion is limited to the theoretical particle and high-energy physics section of *Physics Letters B*. All the information given here corresponds to the papers submitted for publication, not to the published papers proper, appearing together with the contributions in the other fields of *Physics Letters B* and for which the reader may perform a similar statistical analysis wandering through back covers or volume reference lists.

1970. This span of time covers most of the duration of my work as an editor[2] and, as it comes to an end, it seems appropriate to convey to a larger audience information which could be collected and the experience which could thus be gathered in the presently very touchy field of letter publication. In doing so, I have, however, to reluctantly disclose some information which should perhaps have been kept as an editorial secret. I readily apologize to all those who might rightfully feel annoyed at seeing a rejected paper used at building up publicized statistics on field and geographical distributions. Nevertheless, the general patterns which thus emerged show many interesting enough features as to justify publication without hiding too many a pertinent detail, even if some of them may appear as somewhat confidential. Most of this information is actually displayed on Figs. 1 and 2 which show the field distribution of the submitted papers and on Figs. 3, 4 and 5 which exhibit the geographical distribution of the many contributors. I often looked at this editorship as an experiment which, in high-energy physics, often also extends over a two-year period and which eventually leads to plots and histograms which are commented upon, according to prevailing theoretical principles. This paper indeed reflects this point of view.

A large fraction of the information which could be synthesized is presented in Section 1.3. The invariance under time translation or steady slow increases which plots and histograms translate reflects an *a priori* healthy situation and commenting further upon it, if judged necessary, leads to a discussion of the editorial policy which was actually followed during this two-year period. A tentative presentation is therefore made through Section 1.2. These two Sections are indeed presented in a somewhat independent way so that the reader interested in actual figures may readily skip Section 1.2 and go to Section 1.3.

Section 1.3 provides information which indeed goes beyond those proper to the journal. The diagrams and maps show the changing face of high-energy physics as new themes in research start and grow. They display the evolution of lines of approach according to the number of ideas and new results which, as their authors see it, should deserve being quickly brought to the attention of the largest competent audience. They also give a clear idea of the world implantation of this facet of physics, at least as seen from western Europe.

The distribution of the submitted papers depends, of course to some extent, on the general editorial policy which is followed. It appears therefore as interesting after such an experiment to try to abstract some general guiding principles which got into shape while editing the journal. Granting the fact

---

[2] Editing the theoretical high-energy section has so far worked on a turn-over basis. This appears, however, as a mere experimental fact.

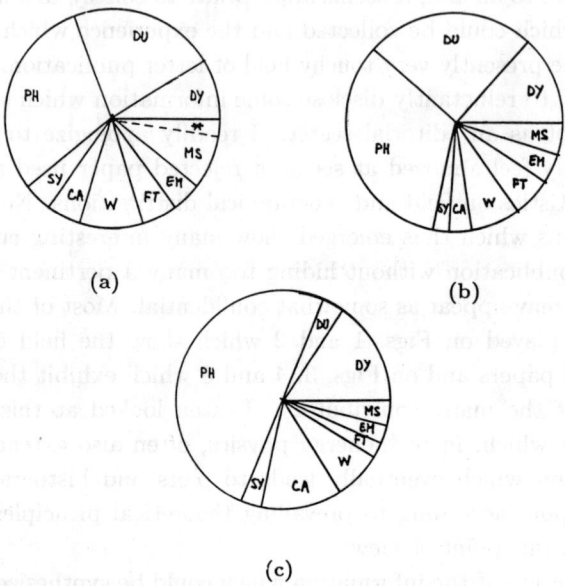

(a)

(b)

(c)

Fig. 1. Field distributions obtained for:
a) the theoretical contributions submitted to *Physics Letters* during 1969 and 1970,
b) the letters actually published during the same period,
c) and the letters published during the preceding two-year period.[5]

The different fields have been separated according to particular labels. Hadron dynamics (DY) refers to letters dealing with relatively formal questions and which do not analyze any specific data, when hadron phenomenology (PH) groups all letters applying theoretical ideas to particular sets of data. Cutting across the two groups we have isolated a duality section, proper to present times. There was no reason, however, to separate out a quark model section which could have been suitable in 1966, or an FESR section suitable in 1968. Phenomenology keeps by essence an ever changing face and this explains its importance in letter publication. We have also separated out symmetries (SY), current algebra (CA), weak interactions (W), field theory (FT), electromagnetic interactions (EM) and finally a miscellaneous section (MS), out of which we may further cut out the few "crackpot" papers (*) which do not propagate to (b). Figure 1 clearly demonstrates how duality, as a specific approach, has channeled a large fraction of recent research in hadron dynamics and phenomenology when current algebra decreased significantly in letter literature. The rise in the yet relatively small number of papers on field theory and electromagnetic interactions reflects recent research in scale invariance and deep inelastic.

As stressed in the text, Fig. 1 demonstrates that the change in trend and fashion is strongly exhibited by the submitted papers and that the editor's action emphasizes it, as expected, but only very moderately so. There is an overall consensus on urgency which indeed prevails.

---

[5]See also: J. D. Jackson — Lund Conference Report. Proceedings of the Lund International Conference on Elementary Particles, G. Von Dardel, Editor.

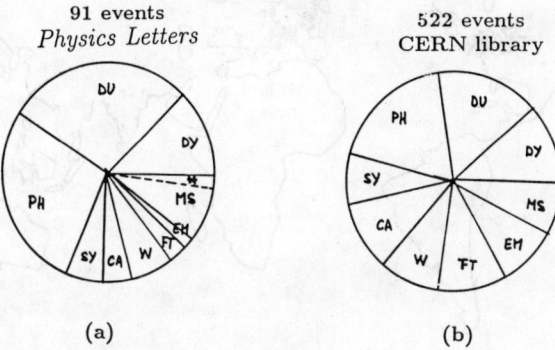

sample period
March–April–May'69

91 events
Physics Letters

522 events
CERN library

(a)   (b)

Fig. 2. Field distributions for the contributions:
a) submitted to *Physics Letters*,
b) and for the preprints received by the CERN library during a sample period of three
 months (March–May 1969).

This clearly shows that the general activity in theoretical physics is much more balanced than
what one would judge from what physicists actually submit to a letter journal. To a large
extent, research sticks to matured fields as well as it switches to new themes. As expected,
research keeps much more a traditional trend than what one would guess by looking at a
letter journal only.

that a letter journal exists, one may try to see why and how it should be used
as to serve best the scientific community and, by the same token, research in
physics. This is outlined in Section 1.2. I should stress, however, that this
represents personal comments which concern the experiment as I followed it.
It does not engage by all means, *Physics Letters B* and even the particular
section of the journal I was in charge of. This should be looked at as a panel
discussion contribution rather than as a full blown analysis of the question.
My approach in this field has been indeed somewhat pragmatic.

1.2. Sketch of A Policy

Why a letter journal?

It is an obvious fact that physics literature is expanding. Journals swell and
split and new periodicals appear. Even in a specific field such as high-energy
and particle physics, it is just impossible to be aware of all the new papers
coming out and for which a worldwide audience is looked for. At present,

Oct 68–June 69

Fig. 3. Geographical distribution of the papers contributed to *Physics Letters B* (high-energy theory) during a two-year period (October 1968–October 1970). Each point represents a submitted paper and, whenever the distribution is dense enough, it is blurred on purpose as to hamper specific assignment. The figure carries actually 250 papers put in chronological order and corresponds to the period October 1968–June 1969.

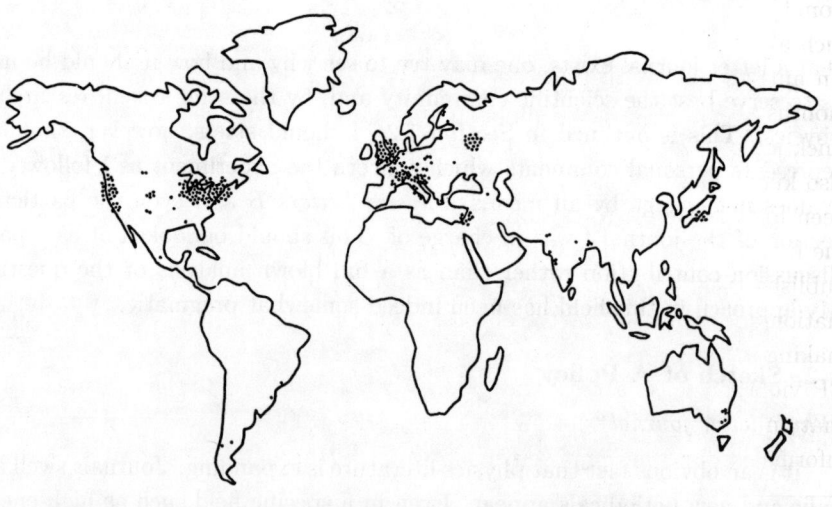

June 69–Feb 70

Fig. 4. Same distribution as the preceding one but for the period June 1969–February 1970.

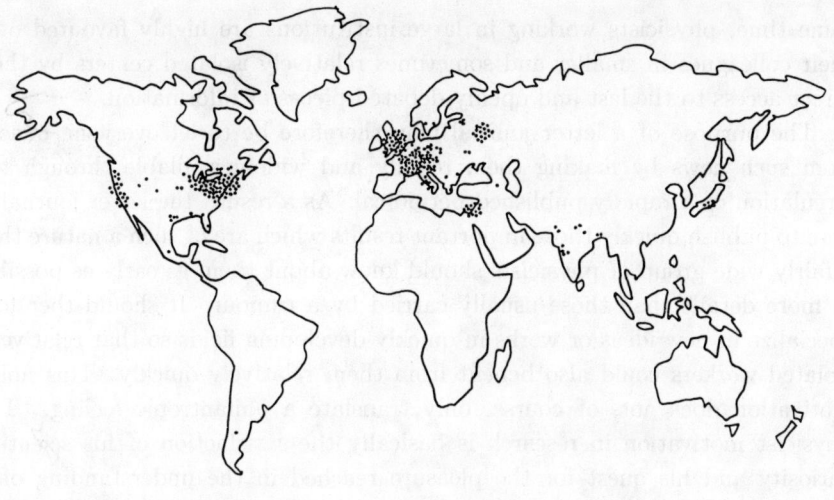

Feb 70–Oct 70

Fig. 5. Same distribution as the preceding one but for the period February 1970–October 1970. The stability of the ratios between the main contributing centers is impressive.*

their number exceeds by far the hundred per week.[3] As a result, many a physicist may feel that even the publication of his work in an international renown journal may not bring him readily the credit which he might rightfully expect from it. Proper attention has often to be called for, through more direct means such as conference communications, seminars, or the circulation of preprints. An author may well think that the eventual publication of his work several months after it has been completed and together with many other papers in a thick journal issue is merely done for archives or later reference purposes. He also knows that some of his colleagues, likely to be interested in it, have long been forced to limit their source of reading information to very few journals, the list of which might not include the specific one which he has selected for publishing his work. It is well-known that the most important pieces of information are spread out by the circulation of preprints and of physicists. Epoch making ideas and great discoveries, such as in recent years, that of the Ω^- and CP violation, have always travelled as lightning. Nevertheless, physics is not that only, and for month to month ideas and results, our standard channels of information have become somewhat obsolete through oversaturation. At the

---

*One may notice the "disappearance" of South America with a last dot from Barloche, where the interference between physics and politic was less strongly felt.

[3]Such a large number is readily met from the union of the CERN and SLAC library accession lists which are distributed weekly.

same time, physicists working in large institutions are highly favoured over their colleagues in smaller and sometimes relatively isolated centers by their direct access to the last and openly debated pieces of information.

The purpose of a letter journal may therefore be to let everyone benefit from such news by making them readily and widely available through the circulation of a rapidly published periodical. As a result, the letter journal is here to publish quickly those important results which are of such a nature that a fairly wide group of physicists should know about them as early as possible in more details than those usually carried by a rumour. It should therefore specialize in new ideas or works in quickly developing fields so that relatively, isolated workers could also benefit from them relatively quickly. This noble motivation does not, of course, only translate a philantropic feeling. The physicist motivation in research is basically the satisfaction of his scientific curiosity and his quest for the pleasure reached in the understanding of a hitherto peculiar phenomenon or effect. It is also the satisfaction achieved through recognition of his merits by his colleagues. Any author is therefore, deeply willing to bring to the widest possible competent audience, and this as rapidly as he can, a new idea which he has just mastered or a new result which he has just obtained. He is therefore often keen on seeing at least a summarized version of his work quickly published and noticed, through its appearance in a letter journal.

Results, the circulation of which is considered as urgent, are generally looked at as important ones even if, as it is sometimes the case, they may turn out to be somewhat misleading or happen to soon become obsolete. Acceptance in a letter journal which to some extent, puts an urgency label on some papers may therefore attach to them a specific prestige not readily granted to standard articles. This assures the journal a steady stream of submitted papers but puts more responsibility upon the editor. Fast publication is possible only if the number of quickly circulated papers is kept relatively small. At the same time, each new issue of the journal attracts immediate and wide interest only because the amount of information which it carries can be quickly assessed and eventually mastered.

If urgent papers may be considered as *a priori* very good, the converse is obviously not true. Nevertheless, in order to prevent the journal from growing much too large, it is mandatory not to take papers which explore in detail, lines of approach which are already somewhat familiar. These papers should be published as regular articles in one of the standard journals and this even if their short length makes them *a priori* acceptable as letters. This is an extremely touchy point since too much prestige seems to be at present attached to letter publication as opposed to publication in one of the standard international journals. One has, however, to admit that a letter journal is well suited at triggering research through the circulation of new ideas and results and that it should therefore stick to this role. New, somewhat technical and interesting results obtained along standard lines of approach stand always a good chance to reach any already interested worker who keeps his eyes open for advances in his own field. Their relatively slow publication will not prevent him from entering a new field of research while still in the making when it could be the case if the rapid circulation of a few letters propagating a new theme or new results had not attracted his attention. It is also often the case that new works in somewhat matured fields highly benefit from being written in more detail and from containing a more adequate discussion than is possible in a letter. In any case, the journal would therefore not meet its purpose if it were not requiring besides interest, some urgency in publishing the contributions which are submitted.

As a result, rejection of a good paper is sometimes necessary. This should, however, not at all be considered as a reflection on the scientific merit of some papers in already well publicized domains. On the contrary, they go beyond what could be said when the corresponding ideas were first explored. Even though they appear as very valuable contributions, there is no clear reason for their fast publication as a letter.

I must say that sticking to such a policy, there were indeed during these two years, very few papers which I was very sorry not to be able to accept. In most cases, these papers were bringing up a new and elegant derivation of an already known result. They were short and a testimony to the author's ingenuity but of course, did not actually bring forward new information. I found out that referees, chosen as to be very knowledgeable in the pertinent fields, were not ready to grant them any urgency at publication. I felt sorry to say "no" when, at the same time, I had to say "yes" for lack of wisdom to some so-called topical papers which quickly turned out to be obsolete.

Since even in doubt about the urgency which should be attached to the publication of his work, many authors try anyway, the rejection rate which has reluctantly to be maintained at a relatively high level. There is an obvious feed-back between the editor's decisions and the authors' attitudes and one should therefore, aim at a relatively steady state where the rejection rate is kept fixed in a natural way[4] and where the probability of getting some violent objection after rejecting a paper is as low as possible. It should be stressed

---

[4]The statistical summary here presented, started being compiled after eighteen months of editorship. The flatness of the curve shown on Fig. 8 is an *a posteriori* finding.

11

that the physicist community is in its majority extremely wise in this respect and that most papers submitted to the letter journal indeed follow the mood and fashion of the time as quickly circulated papers should. This is obviously displayed on Fig. 1 which shows a comparison between the field distribution of research activity in theoretical high-energy physics as obtained from published letters in the 1967–68 period[5] and the one reflected by the input papers in *Physics Letters* in 1969–70. It is obvious that strong interaction physics, with in particular, the duality boom benefited from some transfer from domains which earlier knew a period of glory, such as current algebra. As shown, however, by Fig. 2 which compares the distribution of papers submitted to *Physics Letters* and received as preprints by the CERN library during the same period, the general activity in physics keeps a much more steady feature than what a letter journal may reflect. For the previously mentioned reasons, the difference is even more pronounced when one considers the letters accepted for publication. An even stronger emphasis is put on the newest fields of research but, as clearly demonstrated, the trend is already extremely well defined by the submitted papers.

A touchy question arises with flares of letters in a topical and booming domain. Many examples readily come to mind. During the last couple of years, one may mention the applications of the Veneziano model and the "explanations" of the Serpukhov results. As soon as the initial message is spread over — in general through the publication of a letter — the journal receives a large number of letters pushing the idea further, applying it to other cases or trying something else in the same vein. As they get published, they generate even more letters. The flare would soon saturate the journal if a line was not drawn. This is, however, very difficult since this can only be arbitrarily made. This may even "penalize" some papers compared with others, since many of these letters carry almost the same information which is found out independently and at very similar times. One may then wonder how the letter journal should be a service to the physicist community. Should it serve first the authors and let everyone have his contribution published to the extent that there are good reasons to think that all letters come out of independent works, or should it serve first the reader not imposing him several versions of the same thing, often useful in their slight variety but hardly justifiable in a periodical which is here to quickly carry the newest ideas or very important results. The question is sometimes even more involved due to the fact that the editor cannot always consult the same referees when faced with very similar papers and that, when

---

[5]See also: J. D. Jackson — Lund Conference Report. Proceedings of the Lund International Conference on Elementary Particles, G. Von Dardel, Editor.

doubt starts to sneak in with respect to urgency, a referee might well favour a "yes" when another would have said "no". While an editor, once a letter came from a relatively isolated center, thus acknowledging the "well-known" character of the approach, I used to reject further works which would extend the same theme without bringing in specific new ideas, on the ground that they should be published as regular articles and not as letters. It is easy for an editor to justify any such decision on the ground that one letter in the same vein has already been published. Nevertheless, publication time is of the order of a couple of months, a period during which a flare might well develop. After a two-year experience, I still consider this editorial problem as an unsolved one.

Another point which is worth mentioning in such "memoirs" is the restricted flare which might be generated by a misleading or even wrong paper. The paper having been distributed as a preprint, many a reader who rightfully finds it wrong may think that he should write a letter pointing out why it is wrong. As long as the incriminated paper has not been published, it is not possible to accept the purely negative letter. However, in several cases — one percent of the papers I received — the incriminated paper had already been received and rejected by *Physics Letters*. It was then felt better to readily give a definite "no", motivating it, however, in disclosing what should have perhaps been kept as an editorial secret, *i.e.*, the submission and rejection of the initial paper. The happiest solution is when the first author(s) and his (their) respondent(s) may still find out that there is a good case for a joint and valuable publication. However, it may not work as nicely as this all the time! Leakage of information should be avoided as much as it is possible. In these unfrequent but nevertheless, actual cases there did not *a priori* appear any better way out.

How a letter journal?

The letter journal is successful to the extent that it comes out quickly and faithfully and that it has the widest possible competent audience. Meeting the first condition is the publisher's secret. The second one may be satisfied by the quality of the papers which it carries but also to some extent by the wide distribution of its contributors. In order to come to most research libraries, it should not be too highly specialized. On the other hand, its success in each specific branch of research depends upon an adequate specialization. As a result, the proper structure selected moves with time. As well-known, if *Physics Letters B* maintains nuclear physics and particle physics together, the different editors work on a fairly independent level and, as already stressed,

13

the present discussion applies only to the particular section I was responsible of.

The quality of the accepted papers is naturally a much too subjective affair. Everybody agrees that a letter should be short and explicit. It should of course refer to a new piece of information and some urgency at publication must be attached to it. The first mentioned properties exclude somewhat papers which would benefit from being written in more details and from containing a more adequate discussion than is possible in a letter. Some papers are overcompressed as to meet the length requirement for letters and thus get a chance at being published more quickly. The editor has indeed to make sure that the new paper is carrying information and not basically aimed at providing the reader a deciphering exercise. In order to judge a submitted paper with respect to the later one, the editor heavily relies on the referee advice. Except in obvious cases which involved a tenth, say, of the papers received, no paper was rejected or accepted without my securing first the advice from a referee. I ran on a somewhat artisanal basis and was led to use mostly, the top referee advice readily available at CERN, thanks to the large number of staff members and visitors. In so doing, a decision was most of the time very quickly reached and the acceptance or rejection letter mailed within the week following reception. In some cases, however, the proper referee is to be found only outside, and this involves at least an extra week delay and often more. I am aware of the somewhat biased nature of such a refereeing system and, whenever a contestation arose, comments from an external referee were called for. The probability for getting an objection to a rejection letter turns out to be 0.07 for this two-year period and it was constant. The probability for an angry letter, bitterly challenging the way the whole matter was handled, is smaller by over an order of magnitude. It is of course, easier for an author to switch to another journal than it is for an editor to go against a referee advice and thus running the risk of loosing this referee's help. As some people know it, however, happened a few times, whenever somewhat conflicting referee comments were indeed obtained.

In view of the number of papers received and of the number of rejection letters to write, there is a strong tendency to use standard letters, each one aimed at meeting rather wide specific cases. Nevertheless, one may think that authors who have put much effort and pride in a paper deserve more and expect a somewhat specific and detailed comment. One quickly finds out, however, that a physics discussion which is not pointing out a real mistake but rather weaknesses or misleading points triggers an interesting and even more technical answer from the author. With five rejection letters sent out

14

every week, the editor is immediately overflooded. I have tried a somewhat intermediary system according to which a rejection letter was customized with, in most cases, some technical and pertinent statements inserted in a series of relatively standard sentences borrowed from a prepared stock-pile. As a result, fairly different letters were sent out and much editing time was saved. I have on purpose used many of these sentences in this policy sketch.

I must say that I used to be somewhat flexible with respect to the length without, of course, letting it known. I considered that quality urgency and also clarity should be the basic criteria and that in many cases, a two weeks delay was not worth the somewhat needed shortening. It remains though that a letter should be short and explicit. A short letter is of course, often not self-contained. Nevertheless, it should be kept in mind that if it is quite normal that it may refer to a yet unpublished work for more details, needed clues should be already available in a well publicized form. At the same time, if a summary of a simultaneously published work may be accepted, the same should not hold for a research program.

The geographical distribution of the contributors, which is, as already mentioned, closely connected with the interest raised by the letter journal and in short its success, is described in the next section. This, however, readily brings up two questions. The first point is the vehicular language actually used which is bound to match the vernacular language of the scientific community. It is obvious that letters are published to be quickly noticed by the widest possible audience and as a result, English readily imposes itself. Nevertheless, the linguistic competence of high-energy physicists should not be *a priori* minimized and some liberty should officially prevail as it is the case for *Physics Letters*. I should say, however, that even with this liberty, in two years, as an editor, I received only one letter in French and one in German out of over eight hundred! I am sorry that they could not be accepted. The letter in French actually came from Moscow. The English used by physicists is for many of them not perfect, an understandable fact when one looks at the maps of Figs. 3, 4 and 5 and I would not personally dare to comment about it. Nevertheless, it is obvious that sufficiently broken English may lead to misunderstanding and as a result, some respect for the vehicular language should be enforced. I must say that in most cases, I thought that a few slips could be left rather than asking the author for a better polished version which would have imposed a two-week delay. Only in an extremely few cases did I send back a paper for checking corrections. Authors in doubt of their mastering of the English language should, however, be urged to have their letter checked for general understanding prior to submitting them. The second point is that the journal should offer equal opportunity to all physicists in the world. One may just try to do one's best. When comparing the information here disclosed to the one evidenced by the back covers or reference list of *Physics Letters* during this two-year period, one may, however, realize that the rejection rate has been noticeably low for CERN. I should stress that CERN preprints already have some refereeing system through which the question of submitting or not the paper to a letter journal is also discussed. The general publication policy is also more readily known to physicists there, who, as a result, often prefer to save the pertinent time if acceptance as a letter is not likely.[6] It is, however, not to me to tell whether or not a fair treatment always prevailed. I just hope it was the case. I did not want to use *Physics Letters* as an author but editing it did not leave me too much time to work anyway.* This last remark leads us to the discussion of the number of papers involved.

1.3. World Implantation and Working Rate

A research journal, and in particular, a letter journal which specializes in information which should be quickly propagated to the widest possible audience, seeks a worldwide distribution when the most solid reason for such a distribution is the quality of the papers which it carries but also the actual world distribution of its contributors. It is well-known that research activities exhibit a very strong latitude dependence to the extent that the distributions put on Figs. 3, 4 and 5 where each submitted paper attached a dot to its origin (with some intended blurring effect), grossly match the population density of the Northern temperate zone, they acknowledge the world implantation of the journal and indeed, the type of physics which it serves. The western Europe contribution is of course overemphasized, as it should be for a Europe based journal, but not too strongly so. In any case, the geographical distribution thus obtained follows reasonably well a similar distribution made for the preprints in high-energy theoretical physics received by the CERN library and which exhibits the world activity in this particular field as seen from Europe. Each of the three different maps carries 250 points put in chronological order and all three correspond, in a first approximation, to the same running time. The stability of the ratios between inputs from the main contributors: the United States, Great Britain, the Soviet Union, CERN and Israel is really impressive. They maintain themselves constant within the available statistics. The increase

---

[6]It is probably also the case in the Soviet Union, where postage delays contribute at discouraging any of the least casual attempts. There was also a lengthy approval procedure for papers submitted to a western journal.

*My only "urgent" result during that time was published in *Physical Review Letters*.

of the American contribution, if a significant trend as suggested by Fig. 6, corresponds most likely to a transition period. The geographical distribution of the preprints received by the CERN library, out of which letters constitute of course but a minority, does not show the strong British component found with submitted letters.

A split between three different distributions cut through various times in the year is justified by the lack of any neat seasonal effects. This is illustrated by Fig. 7 which shows the number of papers received per two-month period as a function of time over two years. One may point out "expected" peaks and dips as the one which occurs in the summer of 1970, but none of them is statistically significant. If the last dip is at all mentioned, it is only because it seems to be correlated with other observations. The main effect seems to be a slow increase in input mainly connected, however, with the variation of the American contribution shown on Fig. 6. The stress being put on the interest and urgency of the published papers rather than on their variety and number, a steady regime, or a constant but slow increase which would follow the development of high-energy physics as a whole is to be looked for. The rejection rate plays a great role at maintaining an almost steady input and,

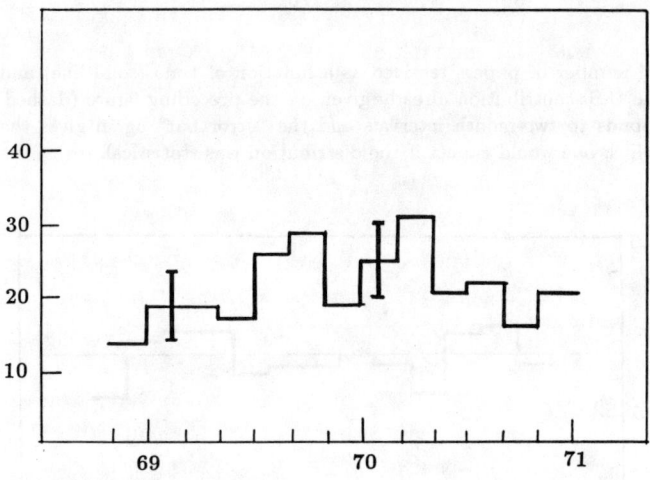

Fig. 6. Number of American papers submitted per two-month periods. The "error bar" indicates the statistical uncertainty which one would expect if the time distribution was statistical. One should not consider too closely the late summer 1970 period since it seems to correspond to a generalized dip in the number of submitted papers. Within the available statistics, one cannot conclude anything definite but it seems that this two-year plot indicates a rise in the number of American contributions.

if everything goes well, also output. Indeed, the rejection rate turns out as very stable over the two-year period considered. One third of the submitted papers gets on the average accepted. This is illustrated by Fig. 8 which gives the rejection rate as a function of time. Short replacement periods by Drs. J. S. Bell

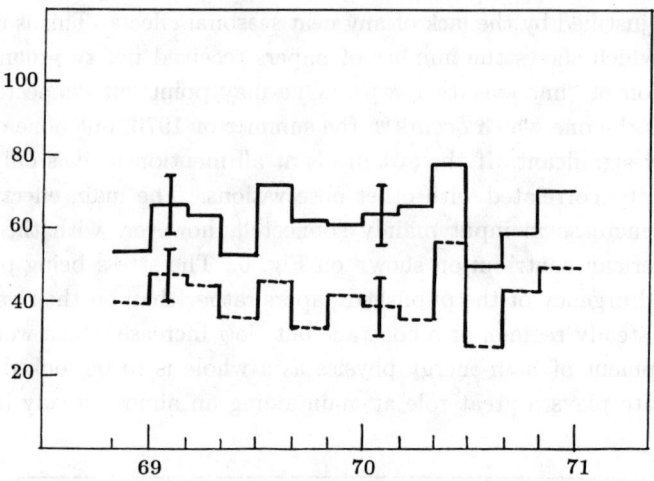

Fig. 7. Total number of papers received as a function of time (solid line) and after subtraction of the U.S. contribution already given on the preceding figure (dashed line). The graph corresponds to two-month intervals and the "error bar" again gives the statistical uncertainty which one would expect if the distribution was statistical.

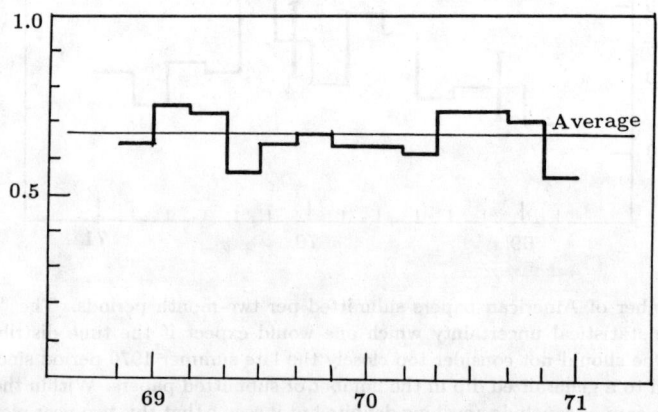

Fig. 8. The rejection rate as a function of time.

(in July 1969) and C. Schmid (in July 1970) gave the same value. The rejection rate does vary with fields of research and naturally it is in the "topical" fields where the input is the largest that the rejection rate is the weakest. The refereeing procedure can only emphasize the preselection of letter publication, as opposed to standard article publication, made by the authors themselves. During these two years, it was of course, weaker in strong interaction dynamics or phenomenology than in more matured or, at present, quiet fields such as weak interactions or symmetries.*

It is therefore, interesting to turn to the field distribution as a function of time. This is illustrated on Fig. 9 which displays the number of papers received in each major field during successive two-month intervals. The actual

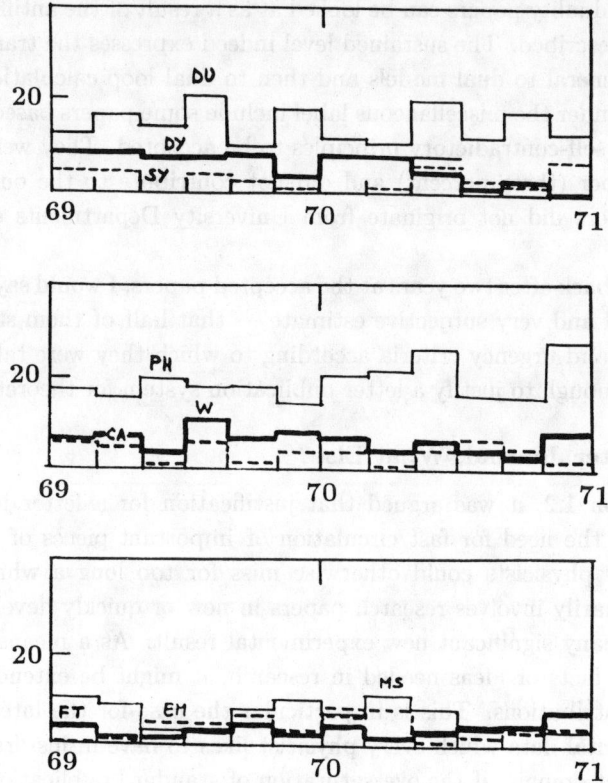

Fig. 9. Time variation of the field distribution of the letters submitted for publication. The fields selected are labelled as indicated on Figs. 1 and 2.

---

*These domains had their revenge in the early seventies!

separation performed is of course, somewhat arbitrary since there are many "overlapping" papers. Nevertheless, the importance of hadron physics is a prominent fact. This merely reflects the fact that new ideas or new models may easily lead to short and explicit letters whereas detailed explorations of already mature fields, if also of top interest, generally call for more detailed and lengthy papers. It is well known that during the last two years, strong interaction physics witnessed with the development of Regge and dual models more "renovation" than the other present domains of particle physics. As already stressed and demonstrated by Fig. 2, the general activity in physics, as reflected by the preprints received by the CERN library, is of course, more traditional than one would conclude from Fig. 9 and much more evenly distributed among the different fields. The not too high and reasonably constant input of the duality papers can be looked at as a result of the antiflare attitude previously described. The sustained level indeed expresses the transition from duality in general to dual models and then to dual loop calculations. Papers referred to under the miscellaneous label include some papers based on too unorthodox or self-contradictory principles to be accepted. They were, however, low in number (three percent) and did not contribute to the output. Most of such papers did not originate from University Departments or Research Institutions.

Looking back after two years at the accepted papers, I would say — but this is a personal and very subjective estimate — that half of them still warrants the interest and urgency criteria according to which they were taken. This is more than enough to justify a letter publication system for theoretical papers.

1.4. A Letter Journal, What Else?

In Section 1.2, it was argued that justification for a letter journal is to be found in the need for fast circulation of important pieces of information which many physicists could otherwise miss for too long a while. This of course, primarily involves research papers in new or quickly developing fields as well as many significant new experimental result. As a means for quickly distributing facts or ideas needed in research, it might be extended to other types of contributions. This is in particular the case for the latest summary of experimental data which every physicist likes to have in his drawers.[7] In a very different domain, if the oversaturation of standard publication procedure did call for separate letter journals many years ago, it seems to call at present for review papers which could act as a guide to the too flourishing literature in any new branch of physics. Any systematic literature survey — and in

---

[7] *Physics Letters* **B33**, 1 (1970).

particular, the modern ones performed with the help of a computer — leads many a physicist to a state where lack of courage might prevail in view of the amount of information actually available for anyone who looks hardly enough for it. As a letter journal was helping with new facts and ideas, it might also help providing guided tours with highly competent guides of newly explored domains or even of themes of research still in the making. It is obvious that there is a need for good review papers which would assess a new field as at least a highly competent author sees it when people have a too strong tendency to write letters instead of more detailed papers! This could of course be dissociated from the letter journal proper.[8] Nevertheless, it was thought better to organize the new review article journal now appearing, *Physics Reports*, as a literature survey of *Physics Letters*, the same set-up being involved in the rapid circulation of important pieces of information and in providing guides to this very information thus made quickly available.

It is well known that summer schools (not all held during the summer) are of a tremendous help at providing their audience with thoroughly discussed material of topical interest and indeed series of lecture notes which come out (perhaps often late) as summer school publication are a good approximation to what one expects from *Physics Reports*. Granting the prominent role which important summer schools have in propagating information among physicists, the announcement of their detailed programs, often known many months in advance but merely advertised through the distribution of posters would gain at being centralized in the letter journal. This is again a tradition which we try to set up at present.

Concluding this review, I would like to express my deep gratitude to all those who were of a tremendous help for me at CERN and elsewhere through their highly competent and rapid refereeing but also secretarial work. I also warmly thank all those in Amsterdam who actually took care of all the published letters. I am also much indebted to Professors M. Veltman, J. Prentki, and J. Hamilton for their most useful advice before I started this experiment and to Professors R. Gatto and E. J. Squires for their promising further continuation. Finally, with the hope that I may leave office without too many more enemies, I would like to thank many authors for their kind understanding and collaboration.

---

[8]In view of the expanding amount of editorial comments associated with the expanding literature, is there already a need for a specialized journal for papers similar to the present one?

RUNNING THE PARTICLE PHYSICS SECTION
OF *PHYSICS REPORTS*, 1970–1985
–1985–

From 1970 to 1985, I was one of the editors of *Physics Reports*. This is a review journal. By the late sixties, physics had already gone through an intense specialization process. This was natural in view of the requirements of research. Yet it was felt that something should be done to break the boundaries which were appearing between subfields. One idea was to provide review articles which would tell the physicist down the hall what was exciting, so much his(her) colleagues a few offices away in the physics department. Another idea was to give the researcher willing to enter a new topical field of research an authoritative survey of the recent topical developments in that domain. The inflation of the literature was already such that it had become difficult to find one's way through the many papers and preprints coming out in any particular subfield. *Physics Reports*, as it took shape, was a compromise trying to meet this two wishes but perhaps, more of the latter one than the former. The journal was designed to cover all fields of physics, but its format, with a separate issue for each of the review papers published, made it a useful working tool for physicists willing to learn a new and important specific development. This was particularly the case for graduate students in domains where books or monographs were not yet available. A large number of people could indeed, thus quickly learn about Gauge theories through the beautiful article of Abers and Lee, published in 1973.

In the case of *Physics Reports*, I took part in the launching of the journal, triggering some of the first articles, published in 1971. I was responsible at first for theoretical particle physics but, after two years, I also took responsibility for experimental particle physics.

The rule was that authors had to be invited but the editor had of course, to be much on the watch to know who was in a position to write a good article provided that he (she) could be convinced to do so. Launching a new journal is a bootstrap operation since, in order to receive good articles, one needs to have published some already. It may work but some people have to be properly enticed to contribute. I was at the origin of almost all of the articles which were published in particle physics but I should say that a few very fine papers in that field also originated. Thanks to the convincing power and good taste of Harry Lipkin.

Running *Physics Reports* in particle physics was a good challenge and a great fun. In that case, one had to compete with the prestigious *Reviews of Modern Physics*. This was even more of a challenge. It was great to receive articles from very famous physicists. It was at least as exciting to trust still unknown young researchers to write review articles which contributed to their budding reputation. Most of them have since become famous physicists. The number of papers coming from America was about equal to that originating from western Europe. There were some very good papers from Russia. It was great to get the first paper from China. All this was very rewarding.

However, time came to leave after fifteen years, which is really a lot for an editor. In 1985, I wrote a paper describing my experience with *Physics Reports*. It was first published (by World Scientific) as a contribution to the G. F. Chew Jubilee Volume, as a farfetched illustration of the Bootstrap idea.

2.1. Introduction

In the course of 1985, I left the Editorial Board of *Physics Reports* after having served on it for nearly 15 years, in effect since the birth of the journal in 1971. *Physics Reports* shares some features with other review journals, yet it is also somewhat unique in several ways. As things developed with time, I practically became responsible for the particle physics section of the journal, which has altogether represented about 30% of the articles so far published. This has been a very interesting experience which I certainly leave with regret at present; nevertheless, I am now no longer able to devote the same attention to it. Furthermore, I have the firm belief that, after these first 15 years, others should take over and bring to the journal new ideas and new contacts. As I leave this editorship, I thought it appropriate to try to summarize in a short essay my present views about *Physics Reports*, bringing together a few facts and a few numbers in which readers of the journal may find interest and even some fun.

24

I started working on *Physics Reports* in 1971, which was also the year during which I left *Physics Letters B*. At that time, I also wrote some memoirs, put together under a very similar title. The style was that of a physics article and it was even accepted for publication, but by a journal which sadly enough never appeared. Attempts at launching a European brother to *Physics Today* have indeed not yet been successful. The *Physics Letters B* memoirs are lost.*
Here are now somewhat similar memoirs covering another editing experiment, this time with *Physics Reports*. They are organized as follows.

Section 2.2 is written somewhat as a testimony, in so far as it is a personally biased historical survey of the development of the journal but, to be honest, mainly of the part for which I was directly responsible. It goes from conception to the present time. It remains at a qualitative level as I try to give the reasons for the different options taken with, occasionally, some interesting crossing of swords on the Editorial Board. Section 2.2 is a more quantitative survey which brings together some data about the particle physics section of *Physics Reports*. Section 2.4 discusses the so-called "Bulk Order Scheme", through which reprints have become financially accessible to many readers. It also discusses the side venture of the reprint volumes. In Section 2.5, I try to conclude with a few general remarks.

2.2. The Conception, Birth and Youth of *Physics Reports*; A Recollection

It was in 1969, at the Lund Conference in Sweden, that I first heard about the intention of North-Holland to publish a journal of review articles which would appear as a special section of *Physics Letters*. I was then the editor for theoretical physics of *Physics Letters B*, and J. Hamilton, who had already discussed the idea with the management of North-Holland, asked me to collaborate in it. The editors of this new section were indeed expected to include a large subset of the editors of *Physics Letters A* and *B*. At that time, the general plan was to put together a journal of review articles aimed at presenting new developments in topical domains of physics, so that non-specialist physicists could thus collect rather detailed information about what was happening in other corners of their discipline. It was recognized that a letter journal such as *Physics Letters* was read mainly by specialists focusing on new papers in their own particular field, readers of the *B* section hardly venturing into the *A* section, and vice versa. It was felt that a new section, the *C* section, was needed in order to propagate efficiently the information between subfields of physics. In a nutshell, the idea was to provide material where the colleague

---

*They are however revived by this book.

25

down the corridor could learn something about what was so interesting at present for his(her) colleagues a few offices away in the Physics Department building.

An original and interesting format was considered for publication. Each article would correspond to a separate issue of the journal, with a typical size of 60–70 printed pages. Issues would be distributed to subscribers and available separately as reprints.

This was certainly a worthwhile endeavour. Nevertheless, I remember having had some reservations at that time. I thought that most of my physicist colleagues would, in general, find little time to enter into details which would take them much beyond the level of information already available in excellent articles currently appearing in *Annual Reviews, Physics Today*, let alone in *Scientific American.* The clever few who knew how to combine ideas originating in different subfields of physics in their own research already knew their way through the original literature, and would not *a priori* resort much to such reviews for general and somewhat delayed, though more easily accessible, information. I was then worried that the need for a new set of such reviews not being clear, it would be too difficult to convince very active physicists who would be the best potential authors, to invest the necessary time and effort in their writing. Credit would be limited and the much-required incentive would not be compelling enough. I now realize that these reservations were to a large extent due to my being a particle physicist at CERN, where emphasis on particle physics is overwhelming and where information is just dumped on you. In any case, I did agree to participate, since I have long incubated an editorial virus, but argued for more technical reviews, which would be rather aimed at dedicated readers willing to enter a new domain of research, and who would thus be ready to invest some effort in their reading.

As far as I was concerned, things remained at this stage for one year. A year later, in 1970, W. Wimmers, director of North-Holland, was still somewhat reluctant to start a new journal: the preprint grinder, which was at that time strongly advocated by some of the editors of *Nuclear Physics.* The main motivations, as I saw it at that time, were as follows. Research physicists had to face an increasing number of preprints. Some were good and even excellent. Some were not very good and sometimes misleading, in particular for the relatively isolated readers. Fishing out the few gems from this extending and unevenly distributed preprint literature was becoming difficult to many, while it was felt that it was at that stage that information had to be collected in order not to miss or catch an exciting new venture too late. This was, of course, particularly the case in theoretical particle physics, which is by essence

26

a frontier domain where promising lines of research may suddenly appear and quickly develop. As the editor for theoretical physics of *Physics Letters B*, I was naturally brought into the debate.

The preprint grinder was, in the mind of its promoters, going to cover some of the new topical developments in physics through a critical analysis of the relevant preprints which had recently become available. A preprint grinder which would thus conduct its readers through the apparently treacherous maze of preprint literature certainly appeared worthwhile. Nevertheless, and probably again because of my personal bias being at CERN, I was not immediately convinced of its usefulness. My main objection, however, was, as I remember it, of a different kind. Having been the close witness of some clashes over priority as editor of *Physics Letters*, and the receiver of many preprints which were never published, I felt that a preprint should not be considered as an actual publication, from which an author could claim some unambiguous credit and priority. Indeed, it is not very infrequent to see several preprints leading eventually to one publication, and preprints leading to none at all. Discussing, let alone mentioning the key points of a preprint in a regular journal, would automatically give it some — I felt undeserved — official status. Furthermore, if negative points were to be raised against a preprint, the author had to be granted some right of response in the journal. This could sometimes lead to the publication of material which would have been better left unnoticed at all. While I certainly realized that the most useful information which one may obtain from preprints is accessible only with great difficulty to many, I felt that giving too much official status to preprints — before the relevant manuscripts had been refereed and accepted for publication — would cause more harm than good.*

I mentioned all this in some detail because *Physics Reports*, as it grew up in the seventies, was actually just a (happy, I should say) compromise between the original proposal of a review journal at a rather general level and the stillborn proposal of a preprint grinder. It was launched as a journal covering rather technical reviews of topical domains of physics. When I say *Physics Reports*, I should rather say those articles in *Physics Reports* covering particle physics since, almost from the start, in the winter of 1970–1971, the different subdisciplines turned out to follow somewhat different editorial policies. The reasons having earlier led to the split between *Physics Letters A* and *Physics Letters B* must have presented an insurmountable problem. Indeed, it later took a lot of goodwill to avoid a split of *Physics Reports* according to the respective specializations of *Physics Letters A* and *B*. This could, happily, be

---

*This is now a logical question with the appearance of electronic data bases for preprints.

27

done, adiabatically only, with the one- and two-diamond markings, which first appeared a few years ago on each issue.

As far as particle physics is concerned, the journal was definitely conceived as a journal of rather technical and topical reviews, providing information to the dedicated reader willing to get involved or, in any case, ready to invest some effort. It thus also appeared as a manageable substitute for the preprint grinder idea. As I remember it, *Physics Reports*, with the format it took, was born in Copenhagen in August 1970 during a meeting with L. Rosenfeld. I had come to Copenhagen with W. Wimmers. At stake was the future of the proposal already discussed a year ago under the name of *Physics Reports*, and the future of the preprint grinder with some of its proponents on the *Nuclear Physics* staff in Copenhagen. Within an afternoon of discussions, the preprint grinder was set aside but, as a compensatory measure, it was decided that *Physics Reports*, at least in theoretical particle physics, where the grinder was almost ready to start its activity, would tend somewhat to the technical side in some of the issues, in order to provide the type of guidance for specialists which the grinder was supposed to give, though in a far more acceptable and general way than that offered by the critical discussion of specific preprints.

I felt very satisfied with the decision and set myself to work. I knew that Gerry Brown and Harry Lipkin, whom I deeply respected as my senior colleagues in this endeavour, were in agreement with this general line of approach. I learned only later that this was, however, not uniformly shared on the Editorial Board which was going to launch the journal. My view of a typical *Physics Reports* article at that time was that of a review which could have corresponded to a series of lecture notes at a Summer School. The Les Houches Summer School, with which I have been closely associated in various capacities for many years, was of course what I had in mind, with its long and thorough survey of topical domains by distinguished physicists. The particular format of *Physics Reports* was especially appropriate. Each review article could be considered as a particular issue of a regular journal (six issues, say, making a volume), but also as a monograph to be used as a small book. This format certainly contributed to the success of the journal.

By the summer of 1971, the Editorial Board met for the first time in Amsterdam. A few articles had already appeared, most of them in theoretical particle physics. This was my first meeting with Dik ter Haar, and I got a cold shower. I had certainly been overenthusiastic and swasivious enough with authors, since particle physics represented 60% of the first five volumes. The emphasis on particle physics and the rather technical nature of the first papers were not at all to ter Haar's liking. I tried to hold my line with my deep belief that the articles which I had commissioned and brought to the journal were at least of high scientific quality, but I still remember ter Haar's quiet voice issuing devastating criticisms. I must say that the discussion was not made any easier by the fact that the two side of *Physical Letters*, *A* and *B*, were actually facing each other in that meeting. With *Physics Letters B*, I was somewhat the involuntary heir to a dispute with *Physics Letters A* which had developed up to an acute level much before my time, and had led to two separate editorial boards. It took the kindness and conciliary attitude of W. Wimmer to reach a "wait and see" conclusion, everyone pledging to do his best for the success of the new journal, which was then seen as an experiment by North-Holland. Its association with *Physics Letters* did allow for an easy try. It went automatically to subscribers, together with *Physics Letters*. One had to wait for their response, which had to carry much weight. We took note of the different opinions expressed on the Editorial Board.

This rather cold meeting was followed by a fairly quiet period. At that time, I had to look for a job. I left *Physics Letters B* after nearly three years as editor for theoretical physics, which had taken a big toll on my time, probably too much for someone with a fixed-term position, and I left CERN for the United States.

My initial flare of enthusiasm at convincing colleagues to write review articles quietened down somewhat. Other editors started bringing numerous articles to the journal. By the end of 1972, when I came back to CERN, the fraction of articles in particle physics had practically reached the level which it was going to keep, with rather small fluctuations, over a full decade, a level of the order of 30%.

By that time, I was convinced that the journal had a clear need to meet and was already doing so to some extent. It could be a substitute for the quickly-diminishing number of postgraduate series of lectures, the world over. After the big boom of the sixties, physics was then experiencing its first difficulties with job openings and funding. This readily led to a decrease of the number of doctoral students and, by the same token, of the number of topical postgraduate courses, once given simultaneously on many campuses. New topics, which used to be taught through special series of lectures, were no longer presented that way in many graduate schools. Nevertheless, there remained a wide but geographically scattered audience for detailed discussions of topical fields. *Physics Reports*, by making available to many a topical series of lectures given at a particular place, or the equivalent thereof compiled as a thorough review by a leading physicist, could supply the needed material to physicists willing to learn about the new developments thus covered. The format of the journal, each article being circulated separately, could help a great deal. This was the direction in which I then wanted to push the journal. The leading figure among such papers is certainly the article *Gauge Theories*, by E. S. Abers and B. W. Lee, which appeared as Volume **9**, No. 1, in 1973, and which has remained for many years the "best-seller" of the bulk order scheme, which I shall discuss later. This is the article through which many physicists, the world over have taught themselves gauge theories over several years. Many other articles, though not quite reaching the same level of fame, have also been extremely useful to many readers, as the sales on the bulk order scheme readily show.

By the mid-seventies, *Physics Reports* had come of age. It was clear that the response met by the new journal was positive enough for it to live on its own. By 1978, formal ties with *Physics Letters* were severed, the *C* appearing next to each number being dropped as a last measure. Subscription to *Physics Reports* could be entered independently of those to *Physics Letters*. The journal had reached a good steady state in terms of subscriptions (close to 1800) and output (about 12 volumes per year). However, the balance between different fields of physics was not yet satisfactory, the usefulness and reputation of the journal among physicists of different fields being rather uneven.

As things went on, I soon took responsibility for almost all the articles published in particle physics; it is worth mentioning at this stage, however, the few but all excellent papers commissioned by H. Lipkin. I started with theoretical high energy physics only in mind but, after a short while, the need for experimental physics reviews was felt and, with the agreement of C. Rubbia,* I started commissioning some and eventually became, to a large extent, the editor in charge of particle physics as a whole. By 1976, the ratio among articles in theoretical particle physics and in experimental physics had reached a value close to three which it has kept since, though of course, with some large fluctuations on a volume-to-volume basis. Practically, all the articles published were commissioned from authors or originated from discussions which I was able to have with the author(s) before work on the article actually started. Indeed, in order to get a good review article on a topical field, one should rely on a very active author who is in essence "overbooked". It follows that launching an article may require a fair amount of convincing, and friendship ties are sometimes instrumental. I take here the opportunity to express my gratitude to the many authors who wrote for *Physics Reports* and who made the journal what it became. Their kind understanding and cooperation was all the time very deeply appreciated. If one wishes to have an important job done well, one has to ask someone who is already too busy. This popular saying took

---

*C. Rubbia was the experimental physicist on the first editorial board.

on a very deep meaning indeed. Following such an attitude may, of course, lead at the same time to some delays, which one has to accept, and I must now express my thanks to the publishers for having accepted proposed dates of submission which turned out to be far too hypothetical. This philosophy may also lead to a rather casual attitude with respect to the actual presentation of the manuscript. One cannot bother a very busy author with too many technical details or favour those who write (or have typed) the better English. I must now express my gratitude to the Desk Editor, K. Korswagen and the people working with her, for the beautiful job which she and they did with manuscripts which happened to be in a really pitiful technical shape when they reached her office. The reader could not see the difference.

As I mentioned already, complete harmony did not always prevail on the Editorial Board whenever it met, which was probably too infrequently to develop efficient contacts. It came close to a fencing match in 1975, when Dik ter Haar and some of the editors associated with the *A* section of *Physics Letters* called for a limitation of the number of papers appearing in particle and nuclear physics. This would have brought them to a frequency level closer to that found with journal publications in physics at large. The editors on the *Physics Letters B* side (as far as fields of interest are concerned) emphasis on some particular domains of physics was very acceptable in so far as the correspondingly more numerous reviews remained of a high scientific standard. Bringing to the journal, a 30% share located almost solely in particle physics, I of course, got a good fraction of the heat. It again took all of W. Wimmers' kind understanding and diplomacy to ease things and avoid a split between two sections along the dividing line between *Physics Letters A* and *B*. All this did eventually lead to some separation, though of a rather adiabatic type (the one- and two-diamond marking). It was recognized that different fields of physics could require different types of coverage and it was thus deemed acceptable that different editors would choose different editorial policies, in so far as each of them was rather closely associated with some particular domain(s), within the same journal. This went a long way towards ending the arguments which had started in 1971 and reached a peak in 1975, as the two sections thereupon went their own way to a large extent. After many years, I think that Dik ter Haar must still think of me as something of a Don Quixote of particle physics. He would certainly be right, since I entered these debates with too much green enthusiasm when more sense of measure would have been appropriate. I think, however, that the journal still needs some strengthening, in quality as well as in quantity in some domains, and in particular some of those associated with the *A* section of *Physics Letters*.

I learned much from these meetings of the Editorial Board, which turned out to be quite fruitful but reaching the proper conclusion certainly made great demands on the patience of W. Wimmers and later of P. Bolman.

In 1981, the two-diamond marking which was covering particle and nuclear physics, where similar editorial policy prevailed, was extended to astrophysics as D. Schramm became editor in charge of that particular domain. *Physics Reports* had by then been recognized as one of the world's leading review journals in physics. It had found its style and its stance.

2.3. *Physics Reports*, A Few Figures

The vitality of a journal can be assessed through the amount of material published per year. Figure 10 shows the number of volumes published every year since 1971. Each volume carries about 400 pages (an average of six issues) and the numbers entered in Fig. 10 are those corresponding to

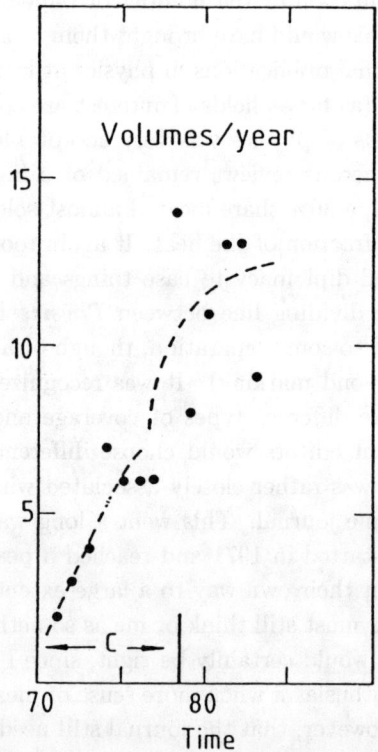

Fig. 10. Number of volumes published per year. The curve translates personal prejudices.

32

the publication dates actually appearing on the covers. Publishers and subscribers may sometimes count differently, considering volumes as attributed to each yearly subscription. The curve corresponds to my personal prejudices. One sees the starting period when the new journal was actually "carried" by *Physics Letters*. Next to this rapid rise, one may see a maturing period, as a relatively high and rather steady state has been reached. In between the two, the "C-marking" (the formal association with *Physics Letters*) had been removed (1978) as the new journal continued on its own. I think that the time evolution should be considered with such theoretical prejudices in mind when venturing an extrapolation. The authors' goodwill, let alone library budgets, may indeed not resist the mere extrapolation of a best-fit straight line giving an average but misleading growth rate!

Contributions in particle physics had an overwhelming role at the very start but, healthily so, they quickly fell in relative, if not in absolute amount. Figure 11 gives the fraction of issues in particle physics per year. By 1973, it had practically reached a constant level with an average value of 1/3 of all the published articles. Reading the relative importance of particle physics from Fig. 11 puts, however, too much emphasis on the first two years during which, while the fraction was high, the global number of published issues remained

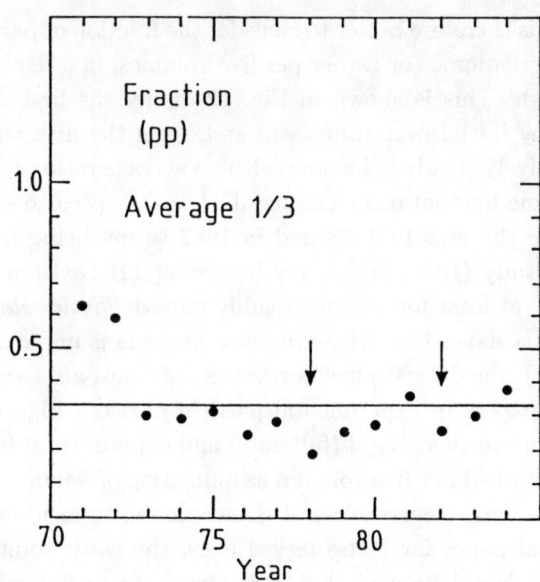

Fig. 11. Fraction of the published issues covering particle physics as a function of time. The straight line corresponds to the average value.

33

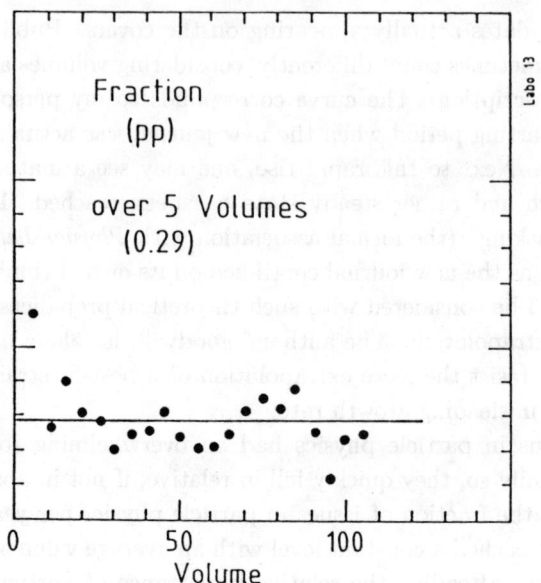

Fig. 12. Fraction of the issues in every five volumes covering particle physics. The straight line corresponds to the average value (0.29).

rather small. It is therefore better to consider the fraction of particle physics by calculating it per volume (or rather per five volumes, in order to partly smear out fluctuations). This is shown in Fig. 12. Only the first data point now acknowledges my initial overenthusiasm and, from the fifth volume onwards, one finds a relatively steady behaviour with an average ratio of 0.29 for particle physics. Are some fluctuations meaningful? I am tempted to say yes, as I can clearly associate the dips in 1978 and in 1982 to my being in charge of the LEP Summer Study (1978) and to my becoming TH Division Leader (1982), activities which, at least for a while, readily moved *Physics Reports* down the priority list. This shows how triggering new projects is important.

I started with theoretical physics reviews only but, after two years, began commissioning papers in experimental particle physics. Figure 13 shows the number of articles in theoretical (full dots) and experimental (open dots) particle physics published per five volumes as a function of volume number. There are large fluctuations. Nevertheless, it is meaningful to mention an average of one experimental paper for 3 theoretical ones, the early points having to be disregarded. It should be said that it is always more difficult to obtain an experimental review article than a theoretical one. When presenting recent and often very topical data, various opinions within the collaboration to which

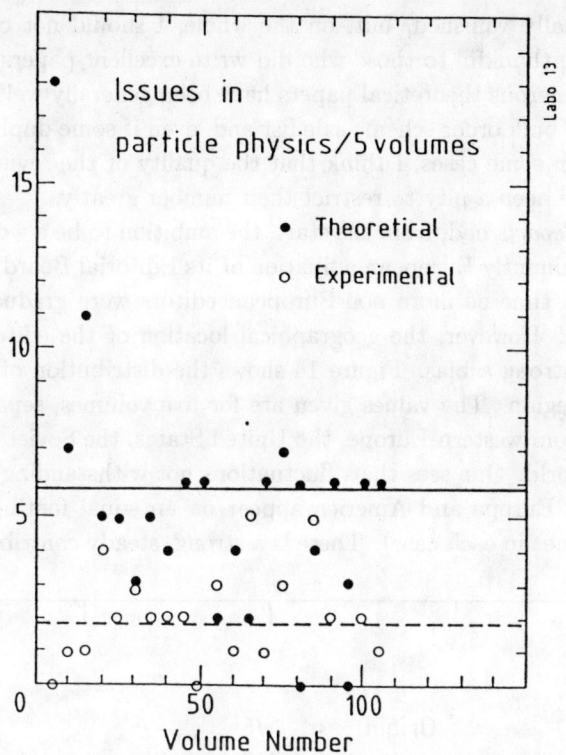

Fig. 13. Number of issues per five volumes in theoretical particle physics (full dots) and experimental particle physics (open dots). The two lines correspond to the average values.

he(she) belongs may easily present touchy problems for the author. Reviewing data from other groups, when one wishes to cover progress in a whole domain of physics, may easily generate still more difficult questions. It is therefore understandable that busy experimentalists are a bit reluctant to embark on an enterprise which may represent some amount of diplomatic work together with a very large quantity of technical work. I am therefore, all the more grateful to all those who contributed to the success of the journal by timely and comprehensive reviews of important developments at SLAC, Fermilab, CERN and DESY over the past ten years. Each review represented a great deal of work, but quickly became a goldmine of information and references. Should one then say that there are perhaps relatively too many theoretical articles? I would certainly have liked to bring to the journal more experimental reviews and I must admit that I failed to be convincing, swasivious or clever enough to obtain timely reviews of important developments. I hopefully followed tracks

35

which eventually vanished, but, on the whole, I should not complain at all and rather be thankful to those who did write excellent papers. On the other hand, the numerous theoretical papers have been generally well received, as is shown by the bulk order scheme sale list and, even if some duplication may be pointed out in some cases, I think that the quality of the reviews justified it. It would have been a pity to restrict their number greatly.

Physics Reports had, from the start, the ambition to be a world journal despite the dominantly European affiliation of its Editorial Board. This actually changed with time as more non-European editors were gradually associated to the Board. However, the geographical location of the editor may not introduce too strong a bias. Figure 14 shows the distribution of the published articles per region. The values given are for five volumes, separating articles originating from western Europe, the United States, the Soviet Union and the rest of the world. One sees that, fluctuations not withstanding, contributions from western Europe and America appear on an equal footing (three issues per five volumes in each case). There is a strong, steady contribution from the

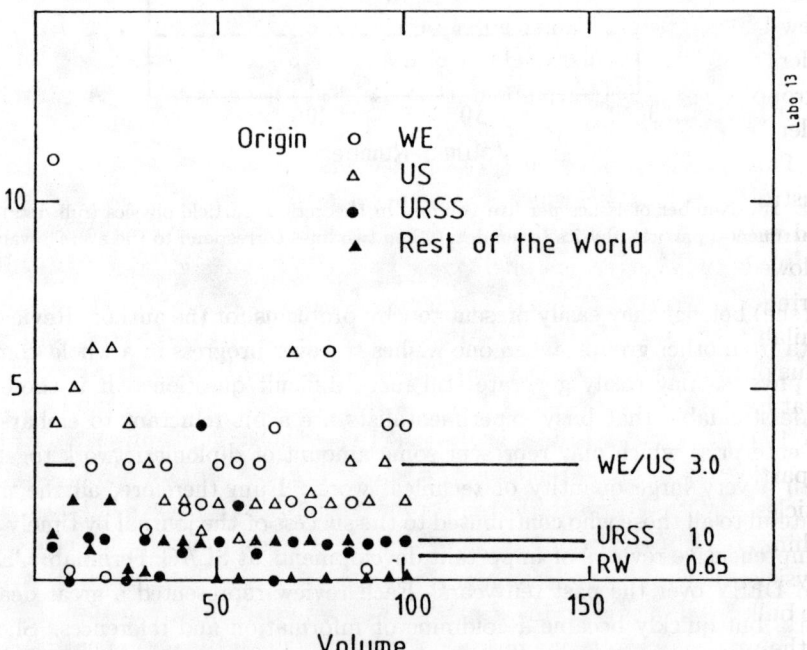

Fig. 14. Regional distribution of the articles in particle physics. Values per five volumes for western Europe (o), the United States (Δ), the Soviet Union (•) and the rest of the world (▲). The lines are average values.

Soviet Union (one issue per five volumes). The contributions from the rest of the world are at a lower level (0.65 per five volumes only). In particle physics, this may however, be deemed a reasonable distribution for the seventies.

2.4. The Bulk Order Scheme

North-Holland publications are rather expensive. However, I have always found a very cooperative attitude when discussing schemes which could ease financially the access to publications for individual buyers. A very successful outcome of such discussions was the "bulk order scheme".

It was recognized that the journal, which over a few years had developed into one of the leading review journals for various branches of physics, had to be regarded primarily as a "library journal" by virtue of the fact that the publisher was obliged to charge a relatively high price for a single issue in view of the individual postage, handling and invoicing charges involved. In order to enable the research worker or graduate student to buy his own copy of a particular report, the following scheme was devised. The publisher agreed to make single issues of *Physics Reports* available to individuals only at a price of a few US dollars (depending on the number of pages), providing that combined orders of at least US \$50 would be placed with them. In effect, this meant that a group of people could combine their separate orders, as long as the collective order amounted to US \$50 or more.

This became effective in 1976 and has been quite successful since. This is illustrated by Fig. 15, which shows the number of issues in particle physics sold every year through the bulk order scheme. Next to a very high point, followed by a relatively low one, both expected of a starting period (1976–1977) during which many readers could choose among the numerous reports already available, one sees a relatively steady rise with time which at present, gives a value of the order of 2000 per year (full dots). Also shown are the sale values of what has been the best-selling article so far: the paper of Abers and Lee, PR 9 No. 1 (1973), which has been at the top of the sale list for many years. Articles in particle physics have represented a practically constant fraction (2/3) of the articles requested through the bulk order scheme. This is shown in Fig. 16 and, I think, reflects the vitality of particle physics, a frontier domain where many physicists are willing to learn new developments. A relatively large fraction of the bulk order sales (Fig. 16) is indeed contributed by a moderate part (Fig. 12) of the total output. If one now considers those papers most in demand (over 50 times per year and over 100 for 1976), one finds about 60 titles, sometimes repetitive and all falling under the general category of "particles and fields". The articles by Wilson and Kogut, PR 12 No. 2 (1974) Politzer, PR 14 No. 4

(1974), Marciano and Pagels, PR <u>36</u> No. 3 (1978), Eguchi, Gilkey and Hanson, PR <u>66</u> No. 6 (1980) and van Nieuwenhuizen, PR <u>68</u> No. 4 (1981) have thus been much in demand, next to the clear first Abers and Lee article. It may then even seem surprising that excellent experimental physics reviews did not appear high on the bulk order scheme list. One may give two reasons for that. The first one is that reviews in experimental physics usually originate from large laboratories with an associated emission of preprints which are even more cheaply available than issues ordered through the bulk order scheme. The second, and more serious one, which is my conclusion from analyzing the titles most in demand, is that the issues which are most frequently requested are those which provide a self-teaching course, which readers have to consult in full and over a reasonable period of time, even scribbling on pages where necessary.

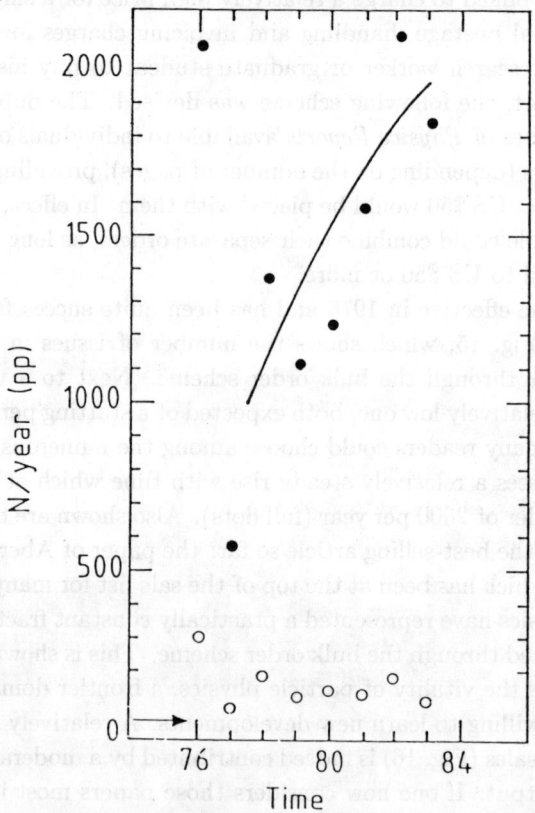

Fig. 15. The number of issues in particle physics sold every year through the bulk order scheme (full dots) and the sale values of the best-seller so far: the Abers and Lee article on gauge theories (open dots).

38

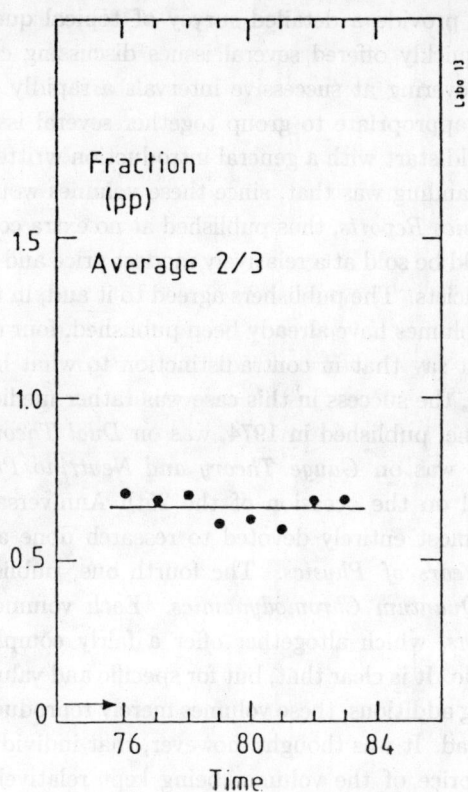

Fig. 16. The fraction of all the issues sold through the bulk order scheme which belong to particle physics.

The best-sellers are actually the working tools. Papers carrying very useful information and of easier reading are merely being photocopied in part from the library shelves and do not appear very high in the bulk order scheme sales.

The bulk order schemes turned out to be very useful, as Fig. 15 shows. The results eased somewhat the discussions on the Editorial Board previously alluded to. The publishers must be thanked for having maintained prices in dollars (the only easy unit) at a time of wild currency fluctuations, and also for having kept them at a level which did not follow the average inflation rate. Services to the physics community carry their importance, and North-Holland has been very cooperative in that matter.

Another venture worth mentioning next to the bulk order scheme is the publication of reprint volumes.

Attempting to provide a detailed survey of topical questions in physics, *Physics Reports* quickly offered several issues discussing closely interrelated development, or covering at successive intervals a rapidly changing field. It was then deemed appropriate to group together several issues into a reprint volume which would start with a general introduction written by a leading expert. The understanding was that, since these volumes would contain mainly reprints from *Physics Reports*, thus published at no extra cost but mere material ones, they could be sold at a relatively modest price and hence be available to individual physicists. The publishers agreed to it and, in the course of time, five such reprint volumes have already been published, four of them in particle physics. One must say that in contradistinction to what happened with the bulk order scheme, the success in this case was rather mediocre.

The first volume, published in 1974, was on *Dual Theory*, the second one published in 1978 was on *Gauge Theory and Neutrino Physics*. The third one was published on the occasion of the 25th Anniversary of CERN and carried articles almost entirely devoted to research done at CERN. Its title is *CERN — 25 Years of Physics*. The fourth one, published in 1982, was on *Perturbative Quantum Chromodynamics*. Each volume contains four to six *Physics Reports*, which altogether offer a fairly complete survey of the corresponding topic. It is clear that, but for specific and valuable introductions and some updating additions, these volumes merely reproduced material which libraries already had. It was thought, however, that individual physicists may be tempted, the price of the volumes being kept relatively low (for North-Holland publications). The sales were, however, not considered encouraging; but I must say that if I had not been deeply involved in the matter, I would not have had, except by chance, any knowledge of their existence. It seems that World Scientific Publishing Company, which has now taken up such reprint publications, has indeed had an encouraging success with the first one which they published: *Collective and Nonlinear Phenomena in Quantum Physics*. This volume puts together the proceedings of the Paris topical meeting of 1975 and that of two specialized sessions of the Les Houches Centre of Physics in 1978 and 1980 respectively. These papers altogether provide a survey of the most interesting developments of common interest to particle and condensed matter physics. One may thus still try another one in 1985, which will then be on *Supersymmetry and Supergravity*.

2.5. A Few Editorial Comments

I much enjoyed my work as an editor and relations with the many authors and with the publishers have been very agreeable, at least as my personal experience goes. While it took a toll on my time, it also helped me much in my work as a research physicist.

Practically, all the articles published in particle physics have been invited, meaning that some discussion took place before the review article was actually written. I already said that, as I see it, good papers require busy authors and a certain amount of convincing may thus be needed, the editor having to make the first approach. Through many contacts, listening to others and eavesdropping, one may however often find the proper time to approach the proper authors. My being at CERN, with extended visits to the large American laboratories, and in particular, to Fermilab in the seventies certainly helped a lot. The steady stream of very good articles from the Soviet Union shows, however, that one can also develop acceptable, though far less effective, contacts through the mail or through its substitutes. I was very happy when I eventually received the first article from China in 1980. Patient efforts were thus rewarded.

Physics Reports articles are refereed, but I felt that refereeing was to be more a service provided to the author(s), assuring them that their paper had been read by a specialist, bringing up in many cases, suggestions and advice for improvements, than a critical examination of the work, leading to its acceptance or rejection. Most authors must indeed be trusted and encouraged to write an article, and they are actually formally invited by the publisher before the paper is written. It would then be somewhat embarrassing not to reach an eventual acceptance. For that reason, I preferred on several occasions to put author and referee openly in contact when the article was still in a draft version, so that the referee's comments could be used fruitfully before the paper was given its final form. I did not receive any bitter complaints, much to the contrary.

Authors invited to write a Physics Reports article are usually well-known specialists and therefore, often senior physicists. I was, however, extremely happy to trust several junior physicists to write a *Physics Reports* before they eventually became well-known. This is a very satisfactory feeling for an editor.

The typical article has been described in Section 2.2. Nevertheless, there have been many atypical ones, sometimes corresponding to experiments which were deemed appropriate, though I must confess that they involved strong personal prejudices which could well not be universally shared.

There was general agreement among the editors that *Physics Reports* should not publish conference proceedings. This should not belong to a regular journal subscribed to on a long-term basis. However, all rules are meant to suffer some exceptions. The only one I considered worthwhile making though, was to take the proceedings of the 1975 topical conference, held at ENS, Paris, and those of the three specialized spring sessions organized by the Les Houches Centre of Physics in 1978, 1980 and 1983, respectively. They all appeared as a set of contributions preceded by a general introduction, bound together as one issue of *Physics Reports*. These meetings covered in great depth the development of the very interesting interface between particle and condensed matter physics. Over the past ten years, this domain has developed quickly with many interesting results. It was thus deemed appropriate to provide a large number of people with reviews of these developments, presented by experts in a commanding, if of course, often in a rather technical way. The response to the bulk order scheme, where the first three already made the best-sellers' list (over 50 requests per year), showed that the exception was indeed worthwhile.

Some issues included contributions by a rather large number of authors. Nevertheless, the organization of the material was highly structured with a general pedagogical introduction, so that the issue could be used as a self-teaching monograph. A few issues covered the research programme of large laboratories rather than physics topics proper. I had mixed feelings about it and the corresponding series has had only three numbers so far. I have to confess that in some cases, I have used the journal to provide a monograph in a domain which I felt deserved coverage while still hardly existing. In such cases, I would ask several experts to write short reviews of different potential facets which would then be put together as an issue of *Physics Reports*. Subjects thus covered included lifetime measurements in the 10^{-13} sec range, the use of synchroton radiation in nuclear physics as it can be considered at LEP, quark matter formation and heavy ion collisions .... I hope that their role in triggering interest may one day justify such ventures. In these cases, I have to be particularly thankful to the authors because some of the times coordination was essential.

It seems that there is a widespread and genuine interest in the recent history of physics, and I was happy to obtain approval for a few *Physics Reports* articles (at most one a year, say, in particle physics), which would review particularly interesting periods in the development of physics. I am extremely happy that E. Amaldi agreed to write the first one. Its title is *From the Discovery of the Neutron to the Discovery of Nuclear Fission*. It appeared in 1984; others will certainly follow.

Several lines of research in theoretical physics have had a rapid and fascinating development, and series of *Physics Reports* could follow them at a yearly interval. This was, for instance, the case for *Dual Theory* in the early seventies and *Quantum Chromodynamics* in the middle and late seventies. It was very interesting to embark on such ventures as an editor, and I am very grateful to the many authors who responded so well.

This has altogether been a very interesting experience and a venture which I am sure others will now develop further and better, with new ideas and much enthusiasm.

PUBLISHING AND EDITING: SOME REFLECTIONS ON EIGHTEEN YEARS AS AN EDITOR
−1990−

Taking into account both my time as an editor of *Physics Letters B*, presented in Section 1, and my time as editor of *Physics Reports*, presented in Section 2, my association with North-Holland lasted for 18 years. This was altogether a long experience. Even though it was limited to particle physics, I could not refrain from developing general views about scientific editing in a more general sense and I was much honoured when the Management of the company invited me to give a talk on the subject in Amsterdam. It was in 1990. This was the occasion to bring together in a more systematic way, many of the ideas presented in Sections 1 and 2 but with the benefit of age and experience.

I very much enjoyed my long collaboration with North-Holland. I much appreciated the mutual trust which developed with the managing editors, the acquisition editors and the desk editors. I became convinced that the editor of a physics journal has better be a bench physicist engaged in research and that the editor has to make up his(her) own judgement from the referee reports and relate by himself to the authors. This is to be preferred to the often seen practice of merely sending to the authors hardly edited referee reports. When a report is negative, this too often prompt the author to try to explain to the editor why the referee has missed the point.

3.1. Introduction

This report presents some personal views about publishing and editing. They are based on a most worthwhile and eighteen year long experience as an Editor, whilst I was primarily a research physicist in a large laboratory.

45

After being an Editor for *Physics Letters B*, I then became an Editor for *Physics Reports*, where I was responsible for most of particle physics, over the period 1971 to 1986. The journal started in 1971 as a special section of *Physics Letters*. It eventually became a separate publication with its own subscription list a few years later.

It is an interesting challenge to try to derive some general statements from this past experience. While some of the remarks presented here should have a more general validity, they apply in the first instance to particle physics where my own expertise and experience is to be found, and some of them may even be limited to that particular domain of research.

During my eighteen years as an Editor, I have been mainly a research theoretical physicist at CERN with employment at a time, and several one month stays afterwards, at Fermilab. My editorial work benefited enormously from the environment of CERN and of Fermilab and from the instrumental help which I could obtain from the many colleagues met at both places. I am also very thankful for the understanding shown by the management of the laboratories where I was employed. I was always left under the impression that they considered this, sometimes time consuming, editorial work as worthwhile. If I had some success in my editorial adventures, it was to a very large extent, thanks to the very active scientific surroundings to which I was constantly exposed. Conversely, the many contacts which I had as an Editor have had definitely a positive impact on my own research, which being mainly phenomenological in character, touches simultaneously many facets of particle physics.

I had to leave my editorial responsibilities when I felt too hard pressed by other commitments, whilst willing to still keep some time for research. By 1982, I became head of the Theoretical Physics Division at CERN. Starting in 1984, I took important responsibilities with the French Physical Society which have now been replaced by similar ones with the European Physical Society. By then, I had also reached an age at which committee membership and often committee chairing became too easily an important fact of life. Finally, I felt that I should leave to others the opportunity to put their own mark on a journal which I had contributed in directing for 15 years. This is a very long time indeed for an Editor! I quit in 1986. So much about my own experience.

My limiting the discussion to particle physics may seem too strong a restriction. However, this is my own domain of research and the only one which I can claim to know very well.

This is a pilot domain in many ways. On the one hand, it is by essence a frontier domain in physics where research emphasis and even the main themes in research change rapidly with time. On the other hand, its rapid internationalization in Europe has been since followed by several other large domains of research, in physics and in other fields. This is a domain in which the North-Holland/Elsevier publications have achieved a clear worldwide leadership.

The strong comeback of European research in particle physics is now widely recognized. It is well known that by the early eighties, Europe had gained worldwide leadership in that domain. At present, CERN is serving half of the High Energy Physics community worldwide. It provides unique research possibilities to about 80% of the particle physicists from the Member States of the Organization and also to 15% of the particle physicists in the United States. It is less widely known that Europe has also taken a leading position worldwide in publishing in that domain of physics.

If one considers the "top" 30 titles of the eighties in particle physics, defined as those which have presently collected over 400 references, one finds that 23 of them have been published in European journals, and North-Holland journals to be more specific! These 23 titles include 12 papers from the United States![1] In the sixties, Europeans were chaffing under the desire to make their results known and acknowledged in America. A publication in an APS journal was carrying some definite prestige. At present, publishing in a European journal may be considered the proper thing to do worldwide.

The success in publishing is clearly correlated to the recognized success in research and the success of CERN has much to do with that of North-Holland in that domain. However, one cannot say that the latter naturally followed the former. They emerged in parallel. It is therefore, worthwhile to try to analyse the different reasons for this success. In scientific publishing as well as in scientific research, success cannot be taken for granted. It can be but passing without ingenuity and hard work.

This report is organized as follows. Sections 3.2, 3.3, and 3.5 cover general questions about publishing and editing. The emergence of commercial publishers in physics is discussed in Section 3.4, whilst the reasons for the success of North-Holland are addressed in Section 3.6. Section 3.7 is about research libraries which are by now, the main customers for periodicals and scientific books.

3.2. Publishing in Physics

One should try to meet three important and general questions

- Why to publish
- Where to publish
- Is there an underlying dynamics in science publishing?

The first two questions are addressed in Sections 3.2 and 3.3. The third one is the object of Section 3.5.

Why does one publish? Physicists are after solving the mysteries of the material world. This is what motivates their effort and hard work. However, they also wish to be properly recognized for doing so. The appreciation of his (her) peers has much to do with the physicist's motivations.

Curiosity and the inborn urge to understand may come first but a scientific discovery should by essence be accepted by all. It should meet the acceptance of peers in order to be deemed valid and worthwhile. Any physicist wishes that others will follow the trail which he (she) has first blazed. There is the wish to collect credit and appreciation for the results obtained or the ideas first put forward.

Any new result has to be made known. Publishing is the normal conclusion of any research. Much value and weight is always put on the free and quick circulation of ideas and results. Feedbacks, criticisms or appreciation by the research community is considered a must. There is also something true with aphorism "publish or perish". Anyone knows that a publication should:

- Announce new results and/or present new ideas.
- Provide detailed enough information that such results can be assessed, at least by specialist readers.
- Put this information in such a form that it can be easily retrieved, consulted and referenced whenever needed and, for that purpose, efficiently displayed and eventually archived.

It also should:

(i) Do it as widely as possible, thus assuring the largest possible impact on the research community.

(ii) Do it as quickly as possible, and this may be very important for collecting credit.

(iii) Do it as cheaply as possible, since this is associated with (i) for well known reasons.

Now the next point is of course, where does one publish?

With respect to point (i), wide circulation, the publishing means is chosen according to the prestige and associated readership of a particular journal. There is something in the aphorism "Bright ideas travel as lightning". Nevertheless, for most results, the publishing means which is used may add an important bonus. We indeed live in a world of information explosion. New

48

pieces of information pour in at a high rate. It is often difficult to distinguish the looked for signal from an often overwhelming noise. Anyone on the watch knows that some channels are better than others. It is very important to entrust any new contribution to one of these better appreciated channels. The researcher is therefore very selective when it comes to choosing a particular journal for a new paper which he (she) naturally considers as being particularly important.

With respect to point (ii), speed is deemed so valuable that it has become usual to issue preprints as a new paper is submitted to publication, and often even before doing so. This can rush the information to colleagues active in the same field of research. By the sixties, this means of rapid communication had become so natural that libraries started to collect preprints and even to archive them at least for a while. Large laboratories (CERN, SLAC, DESY, in particular), which have long been the natural targets in preprint dispatch and could thus receive almost every preprint in particle physics, organized very successful displays of preprints and made their preprint lists widely available on a weekly basis. This became an important research tool.

However, when considering rapid circulation of new results and ideas, preprint circulation cannot act as a substitute for letter journals, which also developed, starting in the late fifties. They meet the demands for speed and visibility.

Letter journals add indeed their prestige to the letters which they carry. It is well known that they have an often tough refereeing system and a strict editorial policy. A publication in a letter journal can thus be efficiently used at claiming credit. Letter journals also allow all those who cannot distribute preprints in large quantities to make their new results and ideas quickly known to the whole research community which, as is well-known, keeps its eyes open on some letter journals.

With respect to point (iii), cost, one has to recognize that the "golden sixties" are over. Research institutions are hard pressed for funds, and research libraries have to live with a budget which cannot meet the expanding literature. This is particularly the case for periodicals. One is on the lookout for efficient but also relatively cheap publication means.

3.3. Publication Means

First, there are *Letter Journals*. The example in our case is *Physics Letters B*. The purpose is the rapid publication of important new results and/or ideas. A successful letter journal can be recognized by the prestige which it has within the scientific community. One does appreciate the speed with which it publishes papers, but first of all, the scientific interest of these papers. Most important is the trust granted to the journal. This is based on a quick and fair refereeing procedure, a clear and accepted editorial policy and the commitment and efficiency of the publisher who produces and distributes it.

Next come regular journals, and the example in our case is now *Nuclear Physics B*. The role is now the detailed presentation of research results. In view of the existence of preprints and letter journals, very few people still rush to the shelves when a new issue appears, to learn about important news. They have usually already come otherwise. Archiving of information and easy referencing is therefore also an important role of journals besides their providing a full presentation of some new pieces of research, with all the details needed, which a letter presentation can often not carry. The author appreciates the fair refereeing procedure. The reader appreciates the selectivity resulting from the editorial policy.

Finally, one has review journals, and the example is now *Physics Reports*. The need for such journals came from their providing much valuable guides to a fast expanding literature and for their providing self-teaching tools in topical new fields, or valuable surveys of key results for the less dedicated reader. Their success reflects the increasing importance of that form of teaching when it comes to advanced subjects. We are no longer in the sixties when specialized series of lectures were simultaneously covering topical domains in many different universities. The author, in particular the young one, is sensitive to the credit associated with a good review. The interest of the reader can be assessed from the number of ordered reprints.

To this list, one should also add:

Conference proceedings. They have become more and more important for the presentation of new and detailed experimental data. Their inborn deadlines force large collaborations to rally around on some presentation. Their invited papers components provide valuable reviews by leading experts.

Summer School proceedings. They have become an important substitute for monographs in topical domains. This is particularly the case for those summer schools which emphasize long series of lectures covering a subject in depth, such as *Les Houches*.

Monographs and books. They have by essence, a rather long term value. They are instrumental for teaching but, in view of the present pace of research, they are somewhat less important than proceedings as research tools.

In this note, I shall first say but a few words on these last three publication means to later concentrate on the first three, namely, letter journals, regular journals and review journals.

With respect to proceedings, one is but the witness of a large inflation fueled by many topical conferences and workshops, by new series of conferences covering new fields while those covering research with a no longer clear specification refuse to die. Yet much of that material is very useful in particle physics. A publisher should be present and selective with the means for being so. One should keep in mind that people like series. It seems better to cover specific domains at regular intervals and carry the proceedings of some particular and regular conference or summer school, than appearing in a spot like way. North-Holland has so far done quite well along such lines.

At present, a potential development of books is associated with the aging of the physics community following massive recruitment in the sixties and a smaller inflow afterwards. There are more and more people at an age where they may wish to write a book rather than keeping full time with full time research. It is therefore, time for quick and selective action, convincing top people to write books in domains where there is a need for a thorough and pedagogical presentation.

I now turn to letter, regular and review journals. The latter two should be kept separate from conference proceedings except in a very few exceptional cases. There is of course, a natural and financial tendency to circulate conference proceedings through the efficient channels of a journal. This is however damaging for the journal. Librarians watch what they buy and one should not sell the proceedings as research material and compel them to buy them as part of a subscription. The corresponding variation in the number of volumes and, accordingly, in the yearly cost of a journal, is also too much at odds with their budget calculations.

While I first thought that some warning was necessary, I realized that North-Holland had already taken the right steps along that line for particle physics, with a new separate proceedings series: the *Nuclear Physics B* proceedings supplements journal. One should eventually offer it as a separate subscription as it became the case for *Physics Reports* initially distributed as *Physics Letters C*.

I already covered some features which are specific to letter journals, regular journals and review journals, respectively and shall from there on concentrate on some global questions common to all.

(i) *Readership.* This is the cardinal feature when choosing a particular journal where to publish. It is associated with the prestige of the journal and the circulation which that implies and the more so when weighted by the attention which it is known to attract when appearing on library display racks.

(ii) *Speed.* This is a clear need which should complement (i). Even though there should be an understandable hierarchy in publication speeds between letter journals, regular journals and review journals, it is each author's wish that his (her) work appears as quickly as possible once it has been submitted and approved for publication.

(iii) *Fair assessment.* This is the key quality which is at the origin of (i). The author should trust the journal. He (she) should feel that the refereeing procedure will be quick and fair and that the Editor will give personal attention should any important problem arise. The reader should trust the journal, knowing that a good refereeing procedure and editorial policy is always at work and implies some peer approval to the scientific quality of the printed material.

Besides that, there is the bootstrap magic. A journal should attract the good papers and it will do so only if it has published some (and no junk!) already. There is no recipe to make it work, besides triggering some "seed" papers to come.

(iv) *Low cost.* Somebody should pay for the publication. Costs could be covered partly through page charge, or, as in the case of North-Holland publications, only through relatively high subscription costs, with no page charge for the author. Since page charges are usually covered by laboratories or contract funds, and libraries have become practically the sole subscribers, the origin of the money is practically the same in both cases, though often budgeted differently. The amount is of real concern, since, as previously said, library budgets are under strong pressure. The number of printed pages per year is expanding. Cost has become an important concern to the author, the reader [and his (her) library] and the publisher.

3.4. The Publisher

Publication is so much the natural outcome of research that it can be considered as part of it. It is also indeed a much needed research tool. Publishing in physics is therefore, a responsibility of the physics community as a whole. It is for those reasons that publication means have traditionally originated from the physics community itself. Physical Societies have long taken an important role in publishing. This is still very much the case. In the United States, publication in physics research (letter journals, regular journals, review journals, conference abstracts) is still dominated by the American Physical Society. In Europe, the IOP (UK) keeps a prominent role in physics though not in particle

52

physics, whereas, in particle physics, the SIF (Italy) long kept an important role with *Il Nuovo Cimento*.

The EPS has recently entered the research publication world with *Europhysics Letters*. There is a long list the world over. Science Academies have also long kept an important role. However, in postwar Europe, one has seen this publishing role partly "delegated" to commercial publishers with the success of North-Holland, Springer-Verlag, ... particularly, noticeable in physics.

Commercial publishers could indeed meet more efficiently and quickly than the traditional means, the need for publishing at the international level which was going with the increasing internationalization of research in Europe. Sometimes new in a field, with new types of publication needed, they could adapt better. They were also driven by the profit to be collected from a fast expanding market of scientific literature with a great need for fast and abundant information whilst physics research was benefiting from increasing budgets. Such a "delegation" of publishing responsibility to commercial publishers can well continue. One of the keywords with the present management of laboratories is "farming out", namely, not to do oneself what industry can do just as well or even better at a given cost. However, one should stress that when this trend first developed in the late fifties, the support of the scientific community was essential.

There was a mutual trust between the world of research and the world of commercial publishing, the latter doing quickly and well what the former was needing. It is this mutual trust which should be maintained. The incentives of the scientist and of the commercial publisher are not always identical. They should of course, try to educate themselves better in the way of the others. Some current has to flow all the time and personal contacts are instrumental.

3.5. The Dynamics of Scientific Publishing

Because of the criteria (i) Readership and (iii) Mutual trust, there is only one winner. The more successful a journal is, the more it attracts the best papers and still more successful it becomes. This sentence with "more" replaced by "less" is also true! This is not specific to journals. It is the case for all commodities. One may be tempted to quote in that context a biological fact "There is only one species per ecological niche." Authors want to have their papers published in the best journals.

The dynamics is similar to the one discussed by C. Marchetti, from the International Agency for Applied System Analysis, when studying the relative roles of energy related commodities (coal, oil, gas, nuclear, ...) as a function of time.[2] This is the same process of market penetration. To a first approximation, it is described by a logistic equation, namely

$$\frac{dx}{dt} = ax(1 - x),$$

where x is the fraction of the market covered by a particular product and dx/dt its variation with time. Once the parameter a is fitted to some data covering a limited time interval during which the product under study takes a reasonable fraction of the market (5 to 10 percent say), one can predict its increasing penetration, building up according to its own strength (the x factor) and to the volume to be filled [the $(1 - x)$ factor]. The time variation is the famous S-shape curve so well-known from the evolution of biological systems. The same applies to journals. They compete for good papers.

A journal is of course, not alone as different species in competition for a niche, or as different commodities in competition for the energy market. There is indeed a system of coupled logistic equations to solve, where x_i is the fraction taken by journal i and where the upper bound for x_i is such that the sum of all x_i's is normalized to one, using the whole available "market" $M(t)$ of good papers, which changes with time.

The key point is that the time variation of some individual x_i can be much faster than that of M, in particular, during fast growth (or fast decline) which are bound to follow a sustained but small growth (or decline). The rise is feeding itself and so is the decline. The whole market is limited. If one journal rises quickly in interest, the others have to decline.

One may conclude that:

(i) A journal cannot aim at a fair and stable share of the good papers in the publishing market, or at a medium x_i. The only relatively stable positions are at the top or at the bottom.

(ii) It should therefore aim at being first and this in an overwhelming way. There is no room for several winners of the same type in an internationalized field.

(iii) It cannot usually pretend being first in all fields and therefore, has to seek or emphasize that specialization in which it can become the best.

(iv) It should be on a constant lookout. Trust and quality are not granted. They have to be fought for. Small newcomers may apparently almost suddenly start their exponential growth. One should keep in mind the S-shape curve.

One may of course, remark that journals almost never die. There is always some dedicated people and some grant type funding to keep them alive.

54

However, such therapeutic struggle does not make sense for a commercial publisher. His publishing has little meaning if his journals are not in demand and a reasonable profit rewards his effort. A cheap journal is not endowed with eternal life. It costs libraries to receive, display and archive a journal and this relatively important cost is much in mind when deciding on which subscription to sever. On the other hand, an expensive journal is always a target for cuts and it has to defend itself all the time according to quality and readership criteria. The consequence of these dynamics — in particle physics, it is clear that *Nuclear Physics B* grew out of (or rather together with) the decline of *Nuovo Cimento A* — is that the publisher should remain in close touch with the physics community.

The few and specialized Editors should preferably be active members of the research community so that responsibilities are clear and strong communication channels are always open. The publisher should understand their academic criteria and not force on to them too many marketing constraints which may lead to only short-range profits. The publisher should help in making the relation between the Editors and the journal staff as easy and efficient as possible. The Editors and the researchers should on the other hand, educate themselves in the ways of the publisher in order to better understand all problems and constraints.

3.6. The Success of North-Holland in Particle Physics

North-Holland journals could fill a need at a particular time when the full international scene became the proper level for publication. Others could not adapt as well to the rapid changes of the sixties and seventies.

North-Holland accepted a strong specialization of its journals, with great success for some which became "top" journals and less success for others. This was a timely process since it came when a strong specialization became a must for research physicists, very few venturing any longer (though sometimes with dividends) into the research literature appearing in another corner of physics. Physics Societies, which feel obliged to follow a broad range approach, have since often fragmented their journals according to specialities.

The con in this specialization, like that evident with *Physics Letters B*, *Nuclear Physics B* and also from the particle physics emphasis in *Physics Reports* is that a strong presence is felt in some particular field(s) only and a successful publication in one field of physics maybe almost unknown in others. This may result in a limited interest from some libraries in research institutions where the field covered is not represented. The pro is that, it directly leads to a personalized editorship which does help in developing the needed trust of the authors. Personalized action has indeed also much to do in bringing some good "seed" papers to the journal.

With specialization, one may also combine quality with size, two parameters which have to be optimized in a different way for the success of a journal, in particular, letter journals which should not deter the reader by a bulky aspect or too much "routine" research. Among other reasons for the success of North-Holland journals, one should of course, mention the excellent publishing technical work, with printing quality, format and speed at publishing, all of high standard. There was also of course, the willingness to risk capital in new and promising ventures. It takes indeed a few years before one can hope that a new journal imposes itself and becomes a successful commercial venture. Sometimes, a field takes years to take off.

North-Holland operates on a "no page charge but high subscription rate" basis. The latter point has often been criticized but one should realize that someone has to pay for the real cost. High subscription rates deter individual subscriptions. The development of North-Holland journals came however, at a time when individual subscriptions would have been on the decline anyway, libraries defining to a large extent the publishing market. In our present world, physicists are no longer used to paying a subscription out of their own money. We live with cheap xeroxing (developed countries) or low salaries (developing countries).

But first of all, and as already stressed, the success of North-Holland publications did require a good relationship between the publishing office and the Editors and more importantly, with the physics community at large.

The help of CERN in providing a large amount of topical and high quality results aimed at European publishing, was instrumental. Its understanding of the importance of editorial and refereeing work by its staff was also an important element. A close contact with and knowledge of the research community is the key element for success in publishing research material.

The impact of a journal has therefore, to be constantly monitored. This can be done with various indicators: subscription level, number of specific orders, citations received by the published articles, and peer assessment. This requires open channels between the journal staff and the research community. It of course, helps when this staff stays long enough on a particular assignment to establish good contacts with the research community which it serves. I was impressed by the many visits the journal senior staff paid to the research institutions. One should also be on the lookout for gradual changes in scope and orientation which always becomes necessary. Here again, there are various means, namely, editorial meetings, expert advice, peer assessment and eavesdropping.

56

What did I appreciate most during my eighteen years as an Editor with North-Holland from 1968 to 1986?

First, the trust put on me and the very good relations which I had with the successive Directors of North-Holland and so many people of their staff. I appreciated the stimulation of editorial meetings. I was always impressed by the efficiency of the company in meeting needs experienced by the research community and I much appreciated the help and understanding which I always found with the desk editors. Of extreme importance also, as I already stressed, was the help which I could find within the research community and this with a base which extended over the whole world of particle physics.

3.7. Physics Libraries

Since libraries have become the main customers of physics journals, a look at present trends in research library operations is deemed appropriate in this survey. In particle physics, a very good library coverage can be estimated to correspond to printruns of about 1200 copies. The economics of the journal should be defined according to that market. I have been closely associated with the CERN library during the 1984–1988 period, when it was part of the TH Division. I could study the motivations and working mechanisms of a large research library. In supervising its running, I benefited from the reading of the little book by Umberto Eco, *De Bibliotheca*.[3] This book makes a sharp distinction between the library of the past and what a present good library should be. I have particularly in mind, the typical library of a research institution where the periodicals budget is much larger than the book budget (a factor 3.5 at CERN).

The library of the past: Accumulated material to preserve it from the damages of time. It considered the user as a hazard and even protected itself against the many unwelcome ones, as illustrated for instance, by the Finis Africae section described in *The Name of the Rose*, also by Umberto Eco.[4] It took pride in its asset much more than in the practical use which was made of it, a use which could often be dangerous to the order of the place and time.

The present successful library should on the contrary be much user oriented and user friendly. It should collect and keep material only in so far as it serves an important role. Book spines are not there for gigantic displays. The user should indeed benefit from his browsing around, which implies that not too much material should be there on display, as a forest hiding some of its own trees. Some selection is necessary. This even means that some weeding has periodically to be done.

The library should know how to bring to the interested user any material

57

known to exist when it is needed, rather than having too many things always on the shelves. The user should be welcomed and helped. His (her) opinion about the usefulness of the library is deemed more important than the total assets as guidelines for policy. At present, a large amount of information has to be computerized in order to be readily available. This in turn implies a lot of indexation work and also much work may be needed making materials available for graphic display on terminal screens spread around the laboratory which the library serves. Friendly and efficient help should be available for retrieval in databases and access to documents. Networking of library catalogues, access to databases and good cooperation are transforming research libraries into information service centres.

All this is quite costly in manpower and the main cost of a good library turns out to be its personnel cost. This fact is too poorly known. At CERN the manpower cost is about three times greater than the material cost and yet users cannot get all the help which they would like to find round the clock. I am convinced that the quality of a library lies with its staff as much as with its periodicals and book content.

It is one thing to get periodicals and books and too many institutions, in particular in developing countries, complain about costs which they can hardly or not meet. However, unless there is a well trained library staff, getting material is not of much use. Indeed, in view of the global cost of periodicals and books, a library has more and more to serve a large number of users. One quick remedy which I see to cure part of the actual hardships in some countries would be to make periodicals available in soft currencies. This is a problem which should be solved as a good gesture to international collaboration.

The question of research libraries and the centralization which they often imply, naturally brings the discussion to that of electronic publishing. Diskettes have certainly become a quick and efficient way to handle information. Their increasing use should become part of the composition procedure and more and more material is going to be prepared and submitted for publication that way. Published material could also be made available to libraries in that form. However, one has to face the fact that processing and reading equipment becomes quickly obsolete. For example, my present 5 year old Macintosh is not capable of reading some of the materials prepared on a more modern (even Apple) equipment. What about materials which would now be presented on punch tapes or cards, which would have been the state of the art 20 years ago! Nobody can read them anymore.

Materials prepared for electronic equipment is wonderful for quick use but runs the risk of being unaccessible to most users a few years later. For example, when I tried to get an old scanning table for the CERN microcosm exhibition in 1989, I discovered that all the CERN ones had long been scrapped! Yet such tables were still the state of the art in the seventies. How can one now study old bubble chamber pictures while so much film is still available! Electronic detectors have, of course, pushed aside Bubble Chambers. What is true for raw data is however, not true for a lot of published material. Access to it should remain easy. One should therefore keep in mind free access to references, even old ones, and the archive purpose of a library. Many old papers still retain a very high value. For that, there is nothing like printed paper and a good pair of eyes, provided that the paper has not become brittle.

While there is certainly a great future in electronic publishing for the fast set-up and circulation of information, standard lasting paper presentation, directly accessible to the unaided reader, should remain a must for time to come.

References

1. *Top 20 of the 1980s*, CERN Courier **30**, 3 (1990).
2. C. Marchetti, *On Energy Systems in Historical Perspective*, 1980 Gregory lectures, IN2P3 report. More involved models of the Volterra type give long term oscillation. We however deal with a time scale such that a simple logistic approach carries the point.
3. U. Eco, *De Bibliotheca*.
4. U. Eco, *The Name of the Rose*.

DE BIBLIOTHECA: AN ESSAY ON PAST AND MODERN PHYSICS RESEARCH LIBRARIES
–1986–

When I started as a research physicists, it was usual for almost everyone to subscribe to some journals. Many journals had two types of distribution. They had library subscribers and individual subscribers. The latter usually benefited from reduced subscription prices. I lived through a time when things have changed considerably. Individual subscriptions to physics journals have been dropping dramatically. This was due in part to the escalating costs of journals as they had to include more and more issues per year in order to face the expanding amount of incoming material. This was also due to the fact that individual had simply to protect themselves against the invasion of paper. Whereas journals were long looked at for new pieces of information, and were considered a must to consultation as quickly as possible, the extensive and systematic circulation of preprints gave them more and more a reference and archive role. In the early sixties, a physicist's office was usually presenting a good array of shelves covered with the green spines of the *Physical Review*. This is very seldom to be seen at present. One may still find some exceptions with letter journals but this is again, quickly disappearing from the laboratory office scene. Physicists are increasingly relying on libraries for their literature needs. Libraries are making preprints available to them, through display and distribution. Journal sales are more and more directed to libraries only. This limits in sales for publishers has increased the cost charged for any subscription. This has further restricted the sales, limiting it to libraries only.

At present, a physics journal with a circulation of fifteen hundred copies has already a very good impact. The "still rich" library sale, which is the easiest goal to reach, corresponds to less than a thousand copies.

Libraries have to buy since they have to provide the available information to their users. But few people understand that the library budgets have to increase accordingly, in order to meet both the rises in cost associated with the expanding literature and also that which goes with the increase in price of each individual journal as its market is restricted to library sales.

For many years, when I was an editor, I could take but satisfaction in the fact that the journals for which I was responsible for had a good library sale. However, it came a time when I had to look at the matter from the library point of view!

It so happens that, when I was Division Leader of the CERN Theory Division, the CERN library was attached to it. I had to step into library matters rather abruptly since it also turned out that, for one year, there was no professional Head of the library. I was thus confronted with publishing questions as seen from a library side.

Scientific libraries are rapidly changing. They are adapting from an old way of life, where they drew their pride from numerous packed shelves of books and periodicals, which users could consult, to a new style of work, where they have to operate in a more user's oriented way, providing information about what is available, but often somewhere else, and whenever requested, bringing that information to the user as quickly as possible. They cannot hope to have everything of interest at hands and have even often to "weed" away seldom used material in order not to be overwhelmed with paper. I tried to summarize my views in a note which borrowed much from a little book by Umberto Eco which I had just read. I took the liberty to take the same title.

Many people complain about the escalating costs of scientific publications. However, a modern library is of little use without competent librarians and these librarians have to meet an increasing number of tasks for which computer literacy is a must. As a result, personnel costs should actually be the main concern when operating a library. When one decides to sever a subscription to a particular journal because of too restrictive budgetary constraints, the saving is actually not so much in the subscription cost but in the manpower saved at not having to order, receive, shelve for display and eventually bind that particular publication.

4.1. The Modern Scientific Library

The title is borrowed from a recent essay by Umberto Eco, the well-known author of *The Name of the Rose*. In this former book, Eco takes the reader through the maze of a mediaeval library. In the latter essay, he spells out basic differences between what an ancient library was and what a modern one should be. Thinking of the CERN library as a modern and specialized one, there are a few ideas to pick up and implement. In so doing, this note, written primarily for the Library Committee, presents some general remarks and a specific proposal.

An ancient library was first of all a depository of knowledge. This knowledge, accumulated in the form of manuscripts, and later books, was considered as having a lasting value, and the aim of the library was, to a large extent, to preserve it for the future while making it available to a selected few only. Knowledge available in a written form, and often conveying wisdom from past centuries, was indeed deemed to be transferable only to specially initiated people! The lay user was considered an embarrassing hazard, in so far as he could provoke the damage or loss of the precious material collected in the library, or could even spread knowledge considered much better kept under lock and key by those who shared it. The use of the library, copying monks not withstanding, was primarily limited to the librarians themselves.

Times have changed, and a good modern library is expected to be user-oriented and even user-friendly. It should make all available knowledge accessible to its users and, whereas the acquisition rate of an ancient library was slow, the volume of new material reaching an efficient new library is very high and appears to be ever-increasing. This new material comes in the form of books, more and more in the form of periodicals, particularly in the case of a scientific library, but also in the form of specialized reports in a yet unpublished form, the most conspicuous components being the preprints.

On the other hand, whereas the materials collected by an ancient library was considered to have a lasting value and was often deemed worth keeping forever, we have little doubt about the transient value of most of the materials collected by a modern scientific library. While it is often of extreme interest when new and topical, the information contained in periodicals and even books is often doomed to medium-term oblivion, the maximum interest of the users always being focused upon relatively recent materials. One may indeed remark that the paper now used by publishers is sometimes of such a quality that decay is almost inevitable within a few decades.

Although the increase in the amount of new materials available each year may lead one to consider new types of carriers such as microfilm or electronic means, it seems that there will be some opposition to their widespread use, in particular, for the latest topical pieces of information. There is indeed, nothing like printed paper as the working tool which one needs in one's own research, so easy and handy to carry around, and there is also nothing like the pleasure of browsing around in a library, stumbling on something interesting and unexpected while looking for something else. For these reasons, paper will always retain some advantage in the foreseeable future.

Let us summarize at this point. A modern scientific library keeps collecting a large amount of new materials. The new acquisitions should be made known as widely as possible to the users and their access to that material should be made quick and easy. Granting that shelf-space is limited and that a large fraction of the material has an important but only passing value, the library should not hesitate to eliminate material which is unlikely to be in any considerable demand. This is the "weeding" procedure.

The CERN "Preprints and Reports" list is a typical example of a service which our library should provide, and it does so in a very efficient way. All users are quickly made aware of the topical preprint information which the library collects for them on a week-to-week basis. This service has recently been developed.

i) The information is made available on the CERN electronic network before the list can be distributed.
ii) The preprints received by SLAC and DESY and not (or not yet) received by CERN are also listed separately. There are relatively few.
iii) While CERN is faithful to the ISIS system, which it developed for its own use starting from a programme implemented by UNESCO and which meets its needs very efficiently, interrogation of SPIRES from CERN will soon be routinely implemented at the library. It will, however, have to be restricted to the morning hours since saturation problems arise as soon as the East Coast wakes up.

4.2. Where Money Becomes a Problem

Information about new preprints, as just mentioned, is complemented by a preprint display every week. However, it is out of the question, for money and manpower reasons, that preprints could be reproduced and distributed, as many users would rightfully wish. Whenever needed, they should be read on the spot, copied in part or in full, and not merely taken away for good, as too many of them are!

It should be stressed that the same applies to a large extent to periodicals and even to books. The rate at which new issues or new titles are coming out, and the prices charged for them, are such that, while the CERN library should have anything of interest in particle physics at the disposal of its users, it simply cannot subscribe for (or buy) as many copies as it could be considered proper for a wide, quick circulation among the interested users. Over the past

64

few years, budgetary constraints have even led to the discontinuation of certain subscriptions and to a halt in buying books of *a priori* secondary interest. Nevertheless, what has thus been saved is mostly shelf-space and manpower. The most interesting periodicals and books — those which we must have — are frequently also the most expensive ones!

One may lament about the high prices charged by certain publishers. Nevertheless, one has to acknowledge the present system whereby people essentially buy for *themselves* only the few items which are indispensable for their work. They mainly consult, and copy in part whatever they need, from library shelves. Publishers therefore, adjust their prices for topical materials to a market limited to some libraries (of the order of 1000, the world over in particle physics), at the high level which follows from this restricted sale. This may be looked upon as a "bootstrap operation", since individual buyers are then deterred by the high prices and leave libraries as the only customers; there is thus no hope for a change.

There are, however, some very worthwhile attempts to bring topical and specialized material within the reach of everyone's purse. Nevertheless, it seems that, this merely enlarges the accessible library market, to the benefit of relatively poor libraries, while it does not cut appreciably into an actual individual market. Individuals do buy, but are extremely selective in their choice, the reason for this, often being shelf-space as well as financial problems.

We therefore, see in the present trend, a situation where a good and efficient library should have at least one [but seldom more] copy(ies) of any interesting new materials at the disposal of its many users. Even this calls for an expanding budget because of the increase in the amount of new materials available each year — this is part of the information explosion — and the correlated increase in price of periodical subscriptions and of book purchases. It also seems that the contemporary presentation of the material — namely, printed paper — is going to last for quite a while.

In view of this, the proper attitude seems to be:

i) to acquire efficiently new and interesting material as it appears. Although the number of copies will have to be limited, the approach at CERN should be of a broad-band nature covering particle physics and the different related fields of science and technology.

ii) To announce efficiently its presence at the library and make consultation as easy as possible for the users. The library, together with the help of the Library Committee, should do its best in that respect, in particular in improving display, access and retrieval.

iii) At the same time, necessary shelf-space should be obtained, in part, through "weeding", materials which is *a priori* no longer of much use being given away and made available, whenever a request is made, through interlibrary loans.

The fact remains that many users may wish (and often need) to have the materials with them and in many cases, the limited number of available copies make loan good, for a very few but impractical for most. One has thus to acknowledge an increasing necessity for copying on the spot, at least part of the consulted materials, according to the urgent needs of the many individual users. So far, copying on the spot has been free of charge, but slowed down through the use of slow machines, in the hope that users would restrain their needs to what was absolutely necessary. The overall cost was charged to the library operational budget. This amounts at present to about 50 KSF per year.

It now seems that, we should rather help copying on the spot, on the grounds that the big increase in the number of CERN users cannot be matched by an increase in the number of copies available for loan, and in the hope that, easy copying will discourage thefts, in particular for the preprints, many of which quickly disappear for good from the display shelf. The slow copying machines should thus be replaced by faster ones. However, a reasonable limit has to be set in each case, and this can only be done if each user feels the related cost, at least semidirectly. The fast machines should therefore, be operated by means of magnetic cards, thus extending the present night operation to all operations. These cards will be made available to users at a cost, which could be borne out by the Divisions, each division secretariat acquiring a number of such cards and distributing them upon request.*

We hope in this way, to keep the new preprint material always available for consultation by everyone while making it individually available to the interested reader. This should ease the present difficult situation with respect to preprints and also to some extent with some periodicals.

It may seem that, we would in this way be charging other Divisions for a service which we have so far provided for free — though to a lesser extent — ourselves, a frequent practice these days. In the present case, it is however, proposed to operate on an almost *quid pro quo* basis. At present, the permanent loan system, whereby Divisions (Groups) can receive on permanent loan copies of specialized books or periodicals for which they have constant use, while being charged the corresponding cost by the library, involves a comparable budget. The invoicing which it implies is nonetheless very cumbersome

---

*This was however not implemented.

66

and, while appearing on library lists, the corresponding material is no longer easily and widely available upon request, in so far as a Division has already paid for its private use. We would therefore, decrease the charge for permanent loan with the money saved on copying, responding to requests for specialized books by Divisions in a much broader way making requested books freely available to them on a long-term basis. We would then give back in one way, what we request in another with a great saving in internal invoicing and an efficient, though reasonable use of copying.

QUALITY CONTROL IN THE ELECTRONIC AGE
−1993−

Most physicists are now equipped with a personal computer and this allows them to receive documents and print them when they so wish. The access to Electronic Mail is already such that, there is a natural tendency to use it at circulating articles. The time delay between the completion of a paper and its availability to a reader is brought down to practically zero. The distribution can be as wide as one may wish. There is no apparent cost to the individual user. Electronic exchange offers the possibility of a quick interaction between the author and the readers. These are prominent advantages and electronic means are already much in use for preprints. In particle physics for instance, Los Alamos is already operating a widely used system. At a less interactive level, some journals are already operating electronically. They accept papers submitted electronically. They also serve libraries electronically. In the latter case, we have essentially the use of a new and powerful technology with important saving in terms of time and manpower. The traditional operation of a journal, with its usual refereeing and editing procedure, is much speeded up but it remains essentially the same.

In the former case, we have a beautiful research tool which allows a community of specialists widely scattered in the world to interact almost as if they were close by. The problem is however, saturation. It is so easy to send information through an electronic network that the system is likely to become quickly overwhelmed with information of a very uneven nature. With such a wide quick and back and forth circulation of results and ideas, it also becomes increasingly difficult to properly assess credit when researchers are, as well known, very keen of receiving the proper credit for their original contributions. The system exists. It is used. One will have to find the proper ways to adapt to it.

In 1993, I was invited to the Conference of the International Federation of Science Editors, held in Santa Maria Imbato, Italy and could express my views on the matter. My main point was that, even though preprints are now becoming widely available electronically, at least in some subfields already, the need for refereed journals remains very much the same. There are at present, many technical questions associated with electronic publishing which are worth addressing and which I could also discuss. This is a question on which the European Physical Society, among others, had recently invested some efforts. I could report on that.

5.1. Introduction

The views presented here, about editing and publishing are based on a long experience in editing and library matters but which has been limited almost entirely to particle physics. They are also not those of a real professional, since my work as an editor extended over many years but always corresponded to a limited part-time activity. Nevertheless, facts and ideas on what took place in the past three decades are used to address present questions in scientific publications as we enter the electronic age. The demands of particle physics in terms of communication are such that, it certainly provides a very good testing ground.

In scientific research, we very much rely on quick and efficient information about what is happening in our field, and we also want to propagate our ideas and results with very little delay and extensive coverage. We are very many with such needs and wishes and, as a result, we witness an explosion of the scientific literature. The number of articles published per year is increasing and the same applies to the number of journals which publish them. The cost of the journals which carry this expanding literature is increasing up to the point that library budgets can no longer follow, and publication times are often considered too long. It has become rather difficult to separate the signal from the noise browsing through the rapidly expanding amount of published material. It is increasingly laborious to be aware of all new results of direct interest, to locate them rapidly, and to access them.

We have so far relied on information on paper. This format is indeed extremely easy to use, once the good paper is in hands. Yet, we increasingly, use computers and word processors in our daily work and it is tempting to operate with electronic means to increase our quick access to new pieces of information and to store and retrieve materials. However, whereas authors, publishers and readers are all already extensively using computers, these computers are often not able to communicate and paper and traditional type setting are still needed

70

intermediaries. Yet, it is felt that the present impressive flow of paper could be advantageously replaced by electronic exchanges, thus cutting delays and costs and easing search, selection and access, in a world where fast information and communication play a key role.

In the coming years, electronic means are certainly going to substitute many of our present paper-based communications. However, one should see this mainly as a change of hardware, when the "software" aspects of publication, which are based on the editing and refereeing procedures, should keep their fundamental role. Whilst we should more easily handle quantity, we should be careful at maintaining quality control. This is this "software" aspect which is mainly covered here, the specific questions raised by electronic publishing being addressed on the way.

5.2. A Personal Experience

Since, these comments are to a large extent based on a personal experience, a few words about myself are in order.

I am a research physicist and most of my research activity has been in theoretical particle physics while working in large laboratories. As president of EPS, I followed with interest, its activities on publication and in particular, its work on electronic publishing carried in contact with several publishers. As head of TH Division at CERN, I was for a while in charge of the CERN library and could also become well-acquainted with this part of physics communication.

I should say that I found it very important for my editorial work to be at the same time, a research physicist and, as such, a full-time member of the physics community. I am inclined to think that this is what an editor in physics should be.

So much about my own experience. It is strongly biased in the direction of particle physics. Yet particle physics is particularly interesting when considering communication in physics as a whole. It is a very competitive domain, with very good worldwide connections. It is a domain, where many new things occur in view of the frontier nature of the field. Satisfying the communication needs of this field, one is likely to meet the demand of many other domains of research.

I lived this editorial activity through a particularly interesting time. It is well-known that, through its cooperative efforts with CERN, western Europe has regained the leadership in particle physics, a leadership which it had long to leave to America. This can be assessed quantitatively through the number of publications weighted by their citations. CERN alone is now serving half of the

71

world community in particle physics and about 5000 physicists fully rely on it for their research. It is less well-known that this come back was associated with a similar success for European based particle physics journals. For instance, if one considers the 30 "top" titles of the eighties, defined as those which collected over 400 citations, 23 of them were published in European journals and 7 in American journals. What is even more important is that, among these 23 papers published in Europe, 12 of them had American authors. In the sixties, European physicists were chafing under the desire to have their results well-recognized in America, through publication in a leading American journal. At present, a publication in a European journal is considered proper worldwide. In particle physics, the journals of North-Holland and Springer now carry a large fraction of the best papers. Success should however, not be considered for granted. It relies on constant hard work and ingenuity and on the proper monitoring of the needs and wishes of the research community.

5.3. Publishing and Editing

Publishing is the normal conclusion of any research. The progress of science is based on the free and quick circulation of ideas and results and publishing them is the standard way to bring them to the attention of the research community. Authors want to bring their new contributions to the attention of their colleagues and catch their interest. They wish to receive appreciation or criticism, but first of all, they want to be noticed. There is much truth in the aphorism "Publish or Perish".

A publication should be efficiently distributed and displayed, easily retrieved and consulted, and properly archived for later use or reference. This should be done as widely as possible, as quickly as possible, and as cheaply as possible since the "golden sixties" are over.

Since many researchers are now spending part of their time watching the screen of their desk top computer, one may think of a situation where they would have, at any instant, and almost immediately, information about all new publications and, when interested, direct access to them. This is in principle, an achievable goal in the electronic age. However, there is so much new material pouring in, that there is a high risk of being just drawn by a flow of information where the signal-to-noise ratio is often not very high. Some tough selection has to be made. One may certainly try to apply some personal choice while browsing through all the available material, or rely on the advice from colleagues. This is a worthwhile approach but it cannot be the standard one. The only practical way out is to watch particularly, some specific channels of communication and to rely on the selection which they are known to imply.

Whilst science is meant at being rational and objective, one cannot avoid biases in the appreciation of new ideas and new results. Yet the scientific community finds it highly worthwhile to rely on some peer assessment. This is where editing and its associated refereeing procedure comes into the picture. There is a need for selective channels, and the more so that selectivity feeds upon itself. If a particular channel is known to be particularly good for its quality and information content, authors will be enticed to rely on it for their important new contributions. The best channel will attract the best papers and can be even more selective and more precious for its information content. The need for this editing and refereeing quality control will certainly not disappear in the electronic age. The dispatching and handling of contributions and the composition and distribution of journals will probably rely more and more on electronic means. The peer assessment provided by editing and refereeing will remain much as it is. For the same reasons, the present separation between different journals and different types of publications will also continue. This is what is meant here, by communication channels.

At present, the standard procedure is to submit a new contribution to a journal which, if it accepts it, will publish it, and thus makes it widely available to the research community.

However, speed is considered as a cardinal element in communication and publication in a journal may easily take several months and often half a year. Ever since the sixties, physicists have therefore, increasingly rely on preprints, distributing copies of their papers to a certain number of colleagues whom they knew to be engaged in the same research. This is even often done before the paper is submitted to a journal and, at the latest, at the same time. This means of quick communication has rapidly become a research tool and some libraries promptly started to collect and display preprints and even to archive them. Preprints are indeed provided with numbers for reference. This communication means became particularly important in particle physics where research is to a large extent concentrated in a few, very large laboratories. At present, CERN, SLAC, and DESY, each widely provide lists of the preprints which they have received each week and display them to the users of the laboratory while also distributing them worldwide.

The support of the material as it transit from the author to the reader is still paper but lists of the incoming and received preprints are now made available electronically as well as on paper. Within a laboratory, which is quite extensive in size, and where some standardization applies, electronic access to the preprints can be provided, thus saving visits to the library and xeroxing there. In this domain, one sees clearly how the paper intermediary may eventually disappear. However, one encounters on the way, a serious standardization problem since most of the present systems are still unable to talk to one another. Another problem is that electronic systems improve all the time and the products of the new, more efficient ones are often unreadable by those of the previous generation. The new systems may also not incorporate the "old fashion" software needed to read languages deemed obsolete. The large centres can adapt quickly but, unless an important translation work is done on a regular basis, the information collected may rapidly become no longer accessible to many. There is a need for a meta-language which would be translatable in a form acceptable to all systems but any meta-language may not resist the test of time.

Despite all these difficulties, one can see a full electronic approach to preprint circulation coming rather quickly into place. After all, aren't preprints supposed to have but a passing value, yielding eventually to the actually published material. Loosing access to preprints after a while should be acceptable.

Electronic means are usually considered as cheap as compared to traditional ones. However, one has to realize that material handled electronically has to be entered in a very careful way into the system. This is often rather costly, in manpower. For example, at CERN, the preprint service provided to the community at large, costs to the laboratory as much as the total of its periodical subscriptions! The electronic choice is therefore, dictated by speed, efficiency, and practicality, more than by cost. The service now provided by large laboratories in particle physics could eventually be transferred to "preprint servers", dispatching selectively to different users in different fields of science, the information which they would centralize. However, somebody would have to pay.

By the late sixties, when preprint circulation had become standard practice, the need for some selective approach to the quickly expending preprint literature was already badly felt. Some proposals were considered, in particular, at North-Holland with which I was already associated, but nothing direct came up. One of the reasons was that, it was not considered proper to give too much credit to preprint "publications" for which there is very little quality control. Any published statement about the quality of a preprint could trigger a polemic exchange not deemed worthwhile. It was felt that a preprint, unrefereed and only selectively accessible, could not be used to claim credit. Preprints were left as a tool to be used with some care. Instead, the idea of developing review journals, providing practical and authoritative surveys of fastly moving topical fields, was followed. This is why *Physics Reports* was launched and it quickly met a good success.

Indeed, the fast and efficient way of bringing a new contribution to the attention of the research community was already provided by letter journals, which started in the late fifties. Letter journals are there to provide a fast (a very few months at most) publication of relatively short papers of particular interest for the new ideas or the new topical results which they carry. This implies a strong selection but, if it is done properly, the journal adds its prestige to the individual articles which it contains. A few letter journals then becomes the favorite places where to look for information. This information is widely and directly available to all researchers. Authors are happy to entrust a letter journal with their best papers. When I was editor of *Physics Letters B* for theoretical physics, the rejection rate was 2/3 but the rebutal rate (authors complaining about the treatment resulting from the first refereeing procedure) was only 7%. The incoming flux of papers was thus, kept at one per day since a tough editorial policy make authors think twice before submitting a paper at the risk of a rejection. This was in the late sixties and early seventies. The flux has increased much since, and three editors have now to share the work of handling all the incoming papers.

A journal widely recognized as good, attracts good papers and it becomes even better. What is therefore, important is to show a clear and fair refereeing procedure and a clear and accepted editorial policy. One has also to entice some well-known colleagues to publish there. It is also necessary that the publisher commits himself to serve quickly and efficiently the research community, whose needs and ways of appreciation may not always tally with short term economical considerations.

The selection provided by good letter journals will always be considered as a precious element by readers and authors alike. The scientific community fully accepts to submit itself to this peer-review procedure. The readers know where to look for most valuable information. The authors draw some definite prestige from the publication and have confidence that their contribution will thus attract more attention. It is said that "good ideas travel as lightning". This is true but for most ideas and results, the support provided by a prestigious journal is a look for help.

What is going to change in the electronic age? Contributions will eventually be submitted electronically to the publishers. Some publishers accept it already, provided that a particular language has been used. The editorial and refereeing work will be handled electronically. This will save time in postal exchanges no longer needed and in the composition of the journal which will proceed electronically from the accepted "manuscripts". Libraries may receive the journal electronically and make prints themselves in a way serving best their display and archiving procedure, but they may also still wish to receive a printed version from the publisher. This full electronic handling will bring some important saving in publication time and publication costs. However, the editing and refereeing procedure will remain much as it is. Here, time is the time of scientists who serve as editors or referees and it can hardly be compressed in the best journals. One has to rely on the expertise and goodwill of the editors and referees and on their dedication to the research community. Individual journals will keep the same "software" character on which their quality is based. The success of a journal depends very much on the fairness, speed and efficiency of the editing and refereeing work. It is my belief that what matters most for an author is the time which elapses between the moment when a paper is submitted and that when the paper is accepted for publication. Despite the fact that they should, of course, be as short as possible, the printing and distribution times are considered secondary. Indeed, once acceptance is obtained, the author's mind can fully switch to other matters. Credit is secured. I also think that it is important that the editor is a person of flesh and blood who could be approached in case of a serious dispute. One should avoid time consuming triangular problems where the author tries to convince the editor that the referee has missed the point. The editor should make up his mind according to the different referee advices which he or she may receive and be able to face the author without screening himself or herself behind anonymous referee reports. The editor should therefore be a competent researcher in the field covered by the journal, granting the fact that competence is often to know whom to trust. The relative accessibility of the editor is in my mind, very important for the success of a journal. At North-Holland, I pushed successfully for the appointment of some American editors, a move which others found unnecessary for a European based journal. This was the only way to win the confidence of many American authors as we eventually did. I always pushed for as few as possible editors so that, each one of them would feel for some time entrusted with an important responsibility, serving a journal, or a section thereof, which he or she would consider as a personal affair. Even though, one hopes that authors will not bother editors too much, the possibility of a relatively easy access to the person in charge should remain open. The editorial and refereeing procedure should be recognized as fair. This implies the possibility to complain and to argue.

Whilst the editor should be known, the referees should remain unknown on a case to case basis. They will provide their critical assessments as long as their names are hidden from the authors. How the protection of the referee, which presently rely on sealed envelopes, will be maintained in the electronic age still requires some thought and adjustments. One has to live with the extreme competitive character of the research community and the more so, when considering original material submitted to a journal while carrying a good fraction of the author's hopes for recognition for that particular work. Some secrecy has to be maintained.

What has just been said for letter journals also applies to a large extent to regular journals. They usually carry longer contributions which deal in greater depth and detail with questions which may have been only briefly presented in letter journals. Though very valuable, the material which they contain may not have the novelty and urgency usually attached with a letter publication. Yet each journal corresponds to a particular selection channel which authors and readers may particularly appreciate. Journals thus very much rely on their editorial and refereeing procedure in achieving success. Since new features are usually first known through preprints and letter journals, the role of regular journals is to some extent more in archiving results presented with enough details and in providing easily accessible information and references. In particle physics, they are seldom used at disclosing ideas and results in the first place. Their editorial composition and distribution will change in the electronic age. Because of their important archiving role, a new important feature will be the full use of their electronic presentation in electronic scans of the literature. A key point however, will be that they will continue to add their respective prestige as journals to the material, thus located in them.

There are many other ways for relatively quick and efficient communication in physics. There are the review journals, the summer school proceedings and the conference proceedings. Here, also electronic means will speed up the editorial composition and cut costs. It should be said however, that a more efficient way to reduce costs on an overall basis would be to be more selective. Conference sponsors should realize that it is often better not to publish proceedings. Some of this material is however, sometimes highly valuable and often consulted. Selectivity is needed. The role of the editor is very important.

It is my experience that the most popular review papers are those on which readers like to scribble as they move along. They like and need to have their own copy for that! With such material often prepared for a wider, less specialized audience, one should also maintain with its full value the open-minded library browsing which so many scientists like to do from time to time. For that, there is nothing like paper! Paper as a support will remain a must for that material, at least at the display and reading stage and each publisher will probably keep providing such paper copies directly.

5.4. The Dynamics of Journals

Electronic publishing will cut several times consuming steps in the present publication procedure. It should increase speed and efficiency and should reduce costs. However, the editing and refereeing procedures which make the quality of a journal, and of any communication channel, should remain much as they are. There is such a vast amount of published material that the special attention brought to certain journals gives an elitist twist to journal publication. It is very important that several journals operate in competition in the same field of research, leaving a choice to the authors and also to the readers if hard-pressed for time. This should apply to journals operating electronically.

Each journal tries to be the best, at least in a particular field. If recognized as the best, it will attract the best papers and remain the best. Subscription to it will remain a must for libraries. We live with hard-pressed library budgets and libraries have more and more to cut on their acquisitions, often following agonizing choices. Electronic acquisitions will not be free! They may even be more demanding on personnel than standard ones. It is not often realized that the main entry on a large scientific library budget is personnel. The library has to be user friendly, often serving a rather large community of users not always well aware of its internal operation. Modern libraries no longer take their pride from impressive arrays of book spines but in making most of the topical and relevant new material available as easily as possible to their users. Users often need guidance and assistance and this is likely to increase in the electronic age. Selecting purchased material, ordering it, receiving it, shelving it, and eventually, archiving or weeding material is quite expensive in manpower and this, whatever the form in which the material comes and is displayed. The main saving in severing a subscription is not so much the actual subscription cost but the personnel cost at handling the subscription and at making good use of it. Even cheap journals, which may operate thanks to various forms of support, are therefore not protected from cuts, and library choices will tend to be more and more selective according to quality. Time is gone when individual subscribers would represent the main customers for a physics journal. Most journals now operate with library sales as their main output. In particle physics, a circulation of 1000 is therefore, already considered good. Journals will have more and more to defend themselves on quality grounds. Quality, and the interest attached to the journal, should not be taken for granted. One should keep a close watch at the reaction of the community which a journal serves. If some authors start to prefer another journal, the effect may quickly snowball for the reasons already mentioned. It is indeed interesting to follow with physics journals, the same ups and down which one finds with the market penetration of commodities where, for instance, the successive leading role of coal, oil and natural gas as a function of time, are well-known. For a particular field, there is always a rising winner when most of the other ones are on the decline. The position of the winner may however, be eventually challenged successfully by a new comer, or an old timer with a new look, which may start from a modest level but climb because of its better appeal to authors and, as a result, to readers. The rise in the relative importance of journal follows the famous S-curve well-known in the case of commodities and also in biology, where there is only one winning species for each ecological niche. If some journals rise in interest, and as a result in circulation, most of the others have to decline. Editors should be on the watch but so should library committees. We need selectivity and this will continue, and most likely increase, in the electronic age as more materials pour in faster. One will be able to handle efficiently a much larger amount of material and do that more quickly but quality will remain a cardinal element. It is sad to see how many journals are reluctant to adapt themselves. They often try to remain alive through various allocated supports more than through a drastic change in editorial policy. At present, it is for instance, sad to see that the many physics journals in eastern and central Europe are reluctant to cooperate and to specialize in order to achieve some noticeable rank among the existing physics journals worldwide. In western Europe, Physical Societies or Academies, which had long been the traditional publishers of scientific journals, were reluctant to admit the new conditions which presented themselves in the fifties. But for the IOPP, they lost ground to commercial publishers who were quicker and more efficient at adapting themselves to a rapidly expanding international market.

It is sad that library budgets are not following the rising amount of scientific publications. This will impose a greater selection whether or not, the publication procedure becomes quickly electronic. The role of the editorial and refereeing procedure which provides the quality control, will become more important for the survival of individual journals. New systems will favour the emergence of new comers but their success will depend more on their editorial "software" than on the hardware used to treat and channel the published material.

5.5. Some Needed Steps Toward Electronic Publishing

The main present problems are with the great variety of systems used, with their rapid evolution with time, and with the very uneven distribution of these systems within the physics community. A working hypothesis may be that, systems will drop in cost as they become more powerful so that one can consider a point of view where most researchers will have at their disposal, systems which can at least communicate among themselves. The recognition of widely agreed standards would help a speedy implementation of electronic means. However, there is already much vested interest in some existing systems which each serve only a fraction of the scientific community and this will have to be overcome. One is after an increased automation in the submission and publication of papers through the use of interconnected computers. At present, computers are already heavily used by authors and publishers but there are few usable direct links. We cannot realistically consider a unique type of computer or even compatible computers. The solution is therefore, with a machine independent standard that is compatible with existing computer hardware and can be used to transmit documents through a heterogeneous network.

Any text is prepared with a specific "mark-up" which specify its organization. Such a mark-up depends on the hardware being used and each one of them is usually not understandable by a computer of a different type. A "meta-language" is needed to define a "mark-up" which should be independent of the system used. The problem is so general, that scientific publications cannot be considered separately. The "meta-language" is likely to be defined on a much wider basis. Such a language has been developed by the International Standards Organization (ISO) and has been available since 1986. It is called The Standard Generalized Mark-up Language (SGML). This meta-language specifies the manner in which the elements of a document should be indicated in the text so that different mark-up systems could recognize what to do and do it. For this, they follow a Document Type Definition (DTD) which may vary according to the type of documents considered. For instance, there will be a rather special one for publication in physics. An application programme is then used to translate the information, thus coded into a form suitable to any particular word processor.

The key advantage of a single metal-language is that, the number of translation programmes increase as the number of incompatible systems only, and not as the square of it, which would be the case if all types of one to one translation had to be made possible. This is important since a recent survey, made by the Publication Committee of EPS in 1988, gave 20 different types of word processors in use in the large system category, and 85 of them in the small system category! This is quite a Babel tower unless some common language, which could act as a unique intermediary would be implemented. Whilst SGLM presently, exists as a contender to provide mutual understanding among different systems, it has still some draw backs. It is not user friendly

80

because of the specifications which it imposes in the preparation of a text. It is also not yet understandable by most of the WYSIWYG (what you see is what you get) word processors, which people like to use.

EPS, in association with the American Association of Publishers, has invested efforts into the implementation of SGLM for physics publications. A DTD taylored to scientific needs has been developed and proposed to ISO, in 1992. If adopted this year, as it is hoped,* it could gain wide approval as standard and be incorporated into the software of all major new word processors. Manufacturers would have a strong incentive at doing so and this could be incorporated in future products. The output of such systems would then be directly useable by publishers following the international standards. One would still need translators between SGML and other systems which are already in partial but extensive use, such as TEX and LATEX. For instance, the American Physical Society accepts texts in TEX and LATEX since 1989 while editors and referees are still contacted with paper. Technical difficulties can always be overcome but one needs some general agreement on what to do, or aim at.

Steps toward the needed compatibility of the different systems have been made. One may anticipate success but implementation still raises many difficulties. This is a question on which the EPS could do, and still does, a very useful work in association with publishing companies.

To conclude, one may say that one can foresee a situation in the relatively near future where scientific publications will be handled electronically all the way, from authors to readers and through the publishers and libraries, paper output being obtained only when deemed appropriate. This should save time and costs and this should allow the efficient handling of the much increased amount of material which is anticipated as the output of scientific research in the years to come. This will give a chance to newcomers to bring to the scientific publishing world new views and new ways to operate. However, this will not modify the nature of the quality control which each publication channel should impose through its editing and refereeing procedure.

I would like to thank J. Sens, member of the EPS Action Committee on Publication, for most valuable information on the present status of electronic publishing.

---

*This has been the case.

Part II

Working with Physical Societies

PHYSICAL SOCIETIES AND PHYSICS COMMUNITIES: A CONTRIBUTION TO *ACHIEVEMENTS IN PHYSICS*, A BOOK IN THE HONOUR OF V. WEISSKOPF
–1993–

I was twice president of a physical society. In the mid-eighties, I was president of the French Physical Society and, in the early nineties, I was president of the European Physical Society. I had therefore, often to face the questions "What is the use of a physical society?" or "Why should anyone join a physical society?". The proper and more meaningful question should actually be "What is a physical society doing?" It is clear that many people have too little perception of the role of such societies. This is not everywhere the case. In the United States, in the United Kingdom, in Germany and in the Netherlands, most physicists join their national society. As physicists, they associate themselves with what they perceive as their natural professional society and often do it as graduate student already. These physical societies have great prestige. However, in most other countries, physical societies are far from meeting a similar success and the European Physical Society is still another matter.

Having been invited to write an essay as part of a book honouring Victor Weisskopf on the occasion of his 85th birthday, I decided to take that opportunity to talk about my experience with physical societies. I must say that, despite all the critical comments which one may hear about learned societies, they remain great tools for many useful tasks. It is my hope that this text will contribute to making their presence better appreciated. I had difficulties and many problems to overcome. I had frustrating times but also many rewarding experiences. It was wonderful to collaborate with colleagues from other branches of physics and also from many different countries and to learn in their ways of approaching problems.

During my time as president of the French Physical Society, I could push forward some collaboration with the German Physical Society which was there to stay. On the French side proper, this was the time of an interesting debate on the new thesis. I was president of the European Physical Society at a most interesting time when one could capitalize on the big thaw in Europe. The imposing barrier which had long stood across the continent, and which the society had made all attempt to ignore despite difficulties which spoke for themselves, was at last long gone. There were great expectations but also important problems for the physical societies in central and eastern Europe and also for those which created themselves in the former Soviet Union. This was also a time when collaboration with the American Physical Society and the Physical Society of Japan developed much. During my time as president of the European Physical Society, I had the great pleasure to sign three joint letters with the president of the American Physical Society, which were simultaneously published in the bulletins of the two societies. One of them was on collaboration between the two societies and in particular, at the speciality division level. The second one was about collaboration on East-West matters. The third one was about large facilities in physics. It was a great pleasure to see the latter two also signed by the president of the Physical Society of Japan, one year later. At that time, a common statement about physics education could also be issued by the three societies.

6.1. The World of Physics

Viki Weisskopf personifies to perfection, the prominent citizen of the world in physics. Everybody knows him. His impressive achievements in our field and also the very important positions which he held on different occasions readily come to mind. On the other hand, his well-known research works and his important mandates are attached to several different locations so that he seems to simultaneously belong to many places, in short to the whole world of physics.

Physics offers an invaluable microcosm in which to nurse ideas that may eventually help foster understanding and collaboration, for we speak a common language and share a common passion. Viki has been instrumental at promoting good ideas and ideals within the physics community, with an enthusiasm and an energy which we have naturally come to associate with his personality. But, he is also among the very few who know how to successfully propagate such ideas and ideals to the rest of the world. He told us that it is a privilege to be a physicist. As he made us feel, it is also a duty with respect to mankind. Viki stands as a model in our efforts to meet that duty.

Physics is a bridge between individuals of different backgrounds and cultures. It also permits precious links between people who may be kept aside for decades by different ideologies and political systems. In recent times, this was particularly the case at the worst of the cold war, when an imposing barrier stood across the European continent. Viki was among those who spare no effort lest communication and a minimum of scientific collaboration between the western and eastern worlds be brought to an end. Now that the barrier is at last long gone, we have to capitalize on these precious links, maintained at a time of sheer adversity, in order to promote in the best possible way, the great spirit of collaboration which is now present on both sides, ready to bloom. Whilst there are great expectations and great hopes, there are also many difficulties to overcome. They speak for themselves. The example set by Viki together with his continuing advice are of a great help and comfort through this new challenging period.

6.2. Physical Societies, Roles, and Goals

The physics communities have several means at their disposal to pursue this endeavour. One of them is the existence of physical societies. The greatest one, which at present stands as a model for all of them, is the American Physical Society. Yet there are many others in the world, some rather large and very active, others thriving to develop, and some merely at the budding stage in those regions where such associations of people were long discouraged.

It seems quite appropriate to dedicate to Viki an essay on the role which physical societies have and should increasingly take in fostering links within the physics community, and helping it to relate better to the rest of the world. In so doing however, the emphasis will be on physical societies in Europe, a subject on which I collected some experience.

Physics relies on individual achievements. It also increasingly relies on collaborations involving large numbers of physicists who have to work together in order to pursue their research. The physics community at large is a needed resonator for individual achievements. It is also a needed sounding board for the selection of the large instruments increasingly required for research and which, in many places, have already to be conceived, built, and exploited at the international level. Viki is a great physicist. He is also a master at bringing people together. This, he did magnificently on many occasions. There was the Los Alamos period which brought together many physicists who had to overcome their many differences in background and style of work to fully join

87

forces in a common endeavour necessitated by the raging war. There was the CERN period, at a time when European physicists, who had been dramatically split asunder by the war, had to realize and appreciate all that they could do together and only together. Many difficulties and prejudices had to be overcome. The successful outcome owes much to Viki's unfaltering efforts.

At present, physical societies offer a channel for an increased cooperation and collaboration among physicists. It is by far not the only one and any channel is of little value if it is not used for deeds. Physical societies provide a technical and legal framework but also, what is perhaps more important, a legitimacy for actions originating at the grass root level. It would be a pity not to use it. The dedicated spirit of a few individuals is always instrumental. Yet sizeable constituencies are also very important. Many of the so-called learned societies should shake up their often elitist and introverted attitude to become invaluable partners to the deciders of scientific policies and to funding agencies, speaking as they may on behalf of large physics communities. This is particularly the case in Europe where overwhelming governmental institutions have often dwarfed national physical societies and where the European Physical Society, EPS, has to maintain a subtle balance between its federating role as a society of National Societies, and its full international mission achieved through its speciality divisions and committees. Yet physics projects are increasingly international in character and the many wishes of a European physics community at large have to be formulated and heard. Deciders and government advisers actually often welcome such a dialogue. Physical Societies could more and more become their natural partners.

Over the past decade, I found myself increasingly involved in such matters being vice-president and president of the French Physical Society, SFP, for a total of three years, from '84 to '86 and leaving this responsibility to be elected on the Executive Committee of EPS and eventually its president for two successive mandates, from '91 to '93. This is the experience, and the perspectives which it offered to me, which I wish to describe in this essay. I met these duties while remaining full time a CERN physicist, where I was head of the Theory Division from '82 and to '88, and the model set by CERN for international collaboration in science, together with the many fruitful contacts which it readily provides, were instrumental in what I could do at the SFP level and later at the EPS level. I dare say that the example of Viki, who is remembered as one of the great Director General of the Organization, was also very much on my mind, and that on many occasions. This was of a great comfort as I did my best to push forward a few new ideas with the constructive criticisms and instrumental help which I could find, first within the SFP, and then within EPS. This was altogether a very interesting and rewarding experience despite some difficulties often encountered. This was a good challenge altogether. Indeed, one often takes for granted in particle physics a drive for action and an international collaboration spirit which is not yet much developed elsewhere and some inertia and some prejudices have to be overcome.

6.3. Physical Societies in Europe

What is the use of a physical society? With my past experience with SFP and EPS I can try to answer from both, a national and international point of view. First, why are physical societies rather special? There are indeed many professional associations, in particular, in the engineering fields and they are often stronger, wealthier, and more efficient than physical societies. However, physical societies try first of all, to promote a field of research and only, as a by-product, the interests of a particular professional community. This make them somewhat special as professional organizations. After all, in most countries in the world, "physicist" is not a recognized profession. In Europe, this is at present, the case only in Britain and in Spain! Elsewhere, one may be an academic or an engineer, but a physicist "no". The qualification does not exist yet in a legal sense. One may also wonder why physics is rather special in that context and why many physical societies are known at all outside of physics? It turns out that physicists are more efficiently and closely organized among themselves than other scientists. Indeed, they altogether represent a rather large and homogeneous population. They are more numerous than the mathematicians or the astronomers and less diversified than the chemists who include many industrial scientists with a great diversity of interests and functions. Physicists are also often involved in large research collaborations and used to discuss and work together not only to compare and assess their individual results, as they always did, but, and increasingly so, to obtain these results in the first place.

6.4. Physical Societies and Learned Societies

A physical society is an example of what is often referred to with the rather pompous name of "learned society". Learned societies have long played an important role in the development of scientific research. This was particularly the case, during the XVIIIth and XIXth centuries, before universities developed what is known today as scientific departments. Learned societies were meeting places for individuals interested in science. They were associations highly conducive to the exchange of information and ideas. They were promoting science and making it better known through lectures, meetings, publications, and the award of prizes. Some learned societies were rather restrictive, with hard won membership, such as some academies. Others were more freely open to anyone with a dedication to science, as many local and national societies still are.

One may illustrate this with the example offered by the "Société de Physique et d'Histoire Naturelle de Genève", which bloomed in a region which Viki also adopted as his own, and where he comes for summer vacations to the benefit of his many friends there. This society played an important role in local but also European physics at a time when the University of Geneva was practically restricted to Theology and Law. It was created a little over 200 years ago around a few prominent scientists including Horace de Saussure. This is to visit it that Alessandro Volta, who became one of his honorary members, made a one week Geneva stop over in 1801, on his way from Pavia to Paris. It is there that he presented his battery for the first time to a scientific circle. The identity between the "electrical fluid" and the "galvanic fluid", which he was advocating, and thus showing at work, was still meeting with many sceptics who were challenging the battery as a chemical source of electricity, considering it rather as a new kind of condensator. Volta could overcome all criticisms but there were many, and it is fascinating to read the detailed account of his Geneva visit which he wrote up in his diary. He came back to Geneva for a much longer visit the following year and, in 1820, the genevese physicist, Gaspard de la Rive had eventually built a very powerful battery with 500 discs, with which he could redo the famous Oersted experiment. The surprising news of this experiment had immediately reached Geneva, which was a great hub for the propagation of scientific knowledge at that time, and thanks to its thriving society.

Arago was present when the famous experiment was redone in Geneva. He got convinced that a deep connection existed between the "electric fluid" and the "magnetic" one. He had however, difficulties convincing the Academy upon his return to Paris. The Academy had the experiment redone. But Ampère was there and he quickly developed his theory of electromagnetism. De la Rive immediately engaged himself in a detailed series of tests of Ampère's theory something for which he said "testing the theories of Monsieur Ampère requires costly instruments and several good craftsmen." Things have hardly changed, haven't they! For a while, Geneva was a prominent hub of activities testing the unified theory of electricity and magnetism, as it later was, a century and a half past that, testing the unified theory of electromagnetism and weak interactions.

The prominent role of learned societies was however, to fade away in front of the development of large universities with strong scientific departments, which started in the middle of the last century. Learned societies had to adapt to a changing world. They did it in a variety of ways and with different successes.

Specialization in sciences became also a way of life. When Volta visited the Society in Geneva he talked about his battery but he had also many learned discussions about mathematics, chemistry, crystallography, physiology, and even on the virtues of mineral waters, with the people there, and they were all almost on the same footing as those on physics. A century later, there was enough within physics alone to fill the columns of specialized journals and the debates of specialized conferences. Physical societies came into being to cater these needs and to provide a sense of unity to the increasing number of physicists in any particular country. They promoted physics research through the award of prizes as academies continued to do so, over a broad range of fields.

In Europe, a new big change occurred after World War II. There was an imposing development of scientific research, in particular in physics, and this research was financed, assessed, and even often fully carried out, by large research organizations. These organizations were first national but, in 1954, with the creation of CERN, large international organizations also came into being. The role of physical societies at promoting physics and at circulating physics results became of a weaker necessity. On the other hand, physics publications became more specialized and organized on a fully international level. In many cases, commercial scientific publishers were quicker and far more efficient at adapting themselves to the needs of an international community. This eliminated to a large extent, the publishing role of many societies but also the revenues which they could have collected from an expanding publishing market. Indeed, publishing is the natural source of income for a thriving physical society which cannot live from membership fees alone. This is indeed the case for the powerful APS.

The different physical societies were also slow to adapt themselves to the increasing international aspect of research. This was the case for publications but also in other domains. There was a need for international conferences and workshops but they first develop independently of the physical societies. Summer schools bloomed but also largely independently of them. The European Physical Society is created in 1968 but it is largely in the wake of the success of CERN, created in 1954. Bernard Gregory, the Director General of CERN, was indeed of a great help in its coming into being.

Yet lost grounds are not there forever and there are many ways in which physical societies can maintain or regain a thriving role and be recognized as most useful institutions, not only inside the physics community but also outside it.

6.5. The Roles of Physical Societies

There is first the conferences. Physicists need conferences to confront and assess their results with a direct contact not provided by publications. Physical societies organize conferences. They have the legitimacy to decide when a conference is useful and trigger, if needed, the interest of funding agencies. They have in their membership, the competence and interest to make good conferences. Such conferences have different roles. Some of them bring together physicists from different specialities. They aim at a wide attendance and try to provide a survey of new results in the whole of physics. Conferences are however, often specialized, covering only one field of physics. They then allow, at regular intervals, to assess progress in one domain and they are often the occasion for the release of new topical results. They also attract a great attendance. Participating in them indeed became almost a must for an active researcher. Yet today, the most popular conferences are highly specialized ones, often referred to as "workshops". They are used at debating in depth, new results among specialists but, first of all, new ideas. Attendance is usually of the order of a hundred people only.

In Europe, conferences of the first kind have still largely remained of a national character. There is indeed, large enough a constituency in many countries to justify a general conference and it can be run in the local language. National physical societies organize them. The EPS also organizes one general conference but only every three years. However, conferences of the second and third kinds have, in most cases, to be international in order to be worthwhile. The EPS is then substituting itself to the national societies in organizing them. Large speciality conferences are thus organized every year, every two years, or every three years, depending on the field covered. They are set up by the relevant speciality division and often alternate with worldwide conferences, also organized on a biannual basis. These conferences are run in English. There are also many conferences of the third kind which are organized whenever it is deemed appropriate. Worth a special mention in that case are the European Research Conferences, ERC, which are similar in character to the Gordon Conferences in America. They started in 1991. In physics, they are organized by EPS in collaboration with the European Science Foundation, with seed funding from the Commission of the European Communities. Two of them took place in '91, nine in '92 and ten are organized in '93. They have been meeting with a great success.

In fields where series of large European conferences started on their own, such as the conferences on particle physics and those on accelerators, the EPS eventually took over their perennial organization through its speciality divisions, and this has been to the satisfaction of everyone.

Beside the conferences, there are the physics prizes. Physicists are motivated by the desire to understand the nature of the material world in its most fundamental aspects. They are also motivated by the appreciation which they wish to find with their peers. Prizes contribute to that. Distinguishing talents and achievements is one of the roles of physical societies.

At present, the EPS awards a prize in Condensed Matter Physics, the Hewlett-Packard prize, and a prize in Particle Physics. Several past laureates of these prizes have since become Nobel laureates.

Another important role of physical societies should be to provide forums for discussions. Long-range research policies and scientific priorities are defined by specialized bodies at governmental level and physical societies long had but a passive attitude in that respect. However, one realizes that deciders would welcome the views of the physics community at large and would prefer not to rely only upon the advice provided by the hierarchies of the research organizations which they control. One finds them willing to appreciate the views of physicists who could express themselves independently of any particular organization. This is of course, a purely advisory role which may involve a lot of work and bear little effect. Yet it is a duty of the physics community to use that channel as much as possible. One still finds at present, some inertia within physical societies. They are reluctant to engage themselves and the more, so that they have to bear the cost of such consultations. Yet they have the competence to do so and should take initiatives which may well result in specific fundings beside the satisfaction found in advising with some response.

Physics research increasingly relies on large equipments which are exploited by a large number of users. In Europe, these large facilities have increasingly to be planned, constructed and exploited on an international basis. How to best choose such equipments in order to use available resources in the most efficient way? There are specialized bodies but here again, an expression of interest from the physics community at large should be extremely useful and in particular, in domains other than particle physics which, in Europe, is already well-organized at the international level. In time of relatively limited funding for basic research, the approval for a new large facility is becoming inconceivable without the existence of a strong constituency of potential users. How to rally such constituency and convene the many concerned physicists to first debate together what they wish most? The EPS should take an increasing role in such prospective studies. It has the legitimacy to call for the discussion meetings through which priorities at the European level could manifest themselves, and through which strong enough constituency of potential users could realize better their common goals. Why should one attach a top priority to a teraflop computer, a neutron spallation source or a tau-charm factory if there is not a large number of potential users ready to fully exploit the new facility? In Europe, the relevant constituencies can only be widely international when deciders still most often, operate at the national levels. They can but welcome some international advice.

There is also the cardinal question of physics education. It is worrisome to see that, in our modern technical world, fewer and fewer young people are taking physics as a major subject in high school and, as a result, are entering university to study physics. This sad trend is of great concern for the whole physics community. This is a major challenge for physical societies. They are indeed in a good position to act. The smaller ones count physics teachers as well as academic physicists among their membership. The larger ones, the members of which are mainly academic physicists or research physicists in large laboratories or in industry, maintain very good contacts with the societies of physics teachers in their own country. Physical societies should do their best at finding ways to improve physics teaching at all levels, including the high school level, and to associate better high school teachers to a thriving physics community. Much remains to be done.

There are also, the often pressing questions of an ethical nature and, in particular, those associated with human rights. These are questions where the relatively close knit society of physicists, which is rather unique at the world level, can make its weight bear to the full. We should use physics to maintain links against national dictatorial tendencies which may sometimes tend to suppress these links. We should use the moral weight of the physics community to speak out whenever it is deemed necessary. Physical societies have good international contacts and means to assess particular situations rather well. People should feel that physical societies can be relied upon. This is a domain where EPS tries to optimize its actions in association with its many National Member Societies. There is the question of physics and industry. There is the question of physics and development. It is a plate full.

Gilberto Bernardini played a key role in the prehistory and early history of EPS. In 1966, when one starts talking about the creation of EPS, he is past president of the Italian Physical Society and the editor of *Nuovo Cimento*. He has also been a CERN director until '63 and he is a member of the CERN Scientific Policy Committee. How does he then, presents the foreseen goal of EPS? He said that:

- EPS should provide a forum for the discussion of subjects of common interest to all European physicists.
- EPS should provide means whereby action can be taken on matters which cannot be handled by national bodies.

They are very broad and ambitious goals. By 1967, things were well under way and one could speak about EPS, a society which was to come into being in the following year. However, a modus vivendi had to be found with the national societies which did not want to be left aside, and the present structure, with its member societies and individual members, was already agreed. How does Bernardini then redefine the goals of EPS. It is interesting to itemize them according to his views and to reflect on what has been achieved since. His main points were:

- To coordinate meetings and conferences. The EPS now does rather well and it actually organizes most of the major European conferences in physics.
- To act as an international forum. This works to some extent but, as previously said, much more could and should be done. The EPS is a tool through the legal and technical framework which it provides. Yet it is far from being used to the full.
- To coordinate European publications. Bernardini thought that the EPS could help the emergence of very strong European journals in different fields of physics and also provide the board of editors and the panels of referees. He thought that it should also encourage review journals as an antidote to extreme specialization. A little has been done. We are still very far from this ambitious but most natural goal. One of the successes worth mentioning is the launching, in 1985, of *Europhysics Letters*, which, like a phoenix, sprang out of the ashes of *Journal de Physique-Letters* and of *Lettere al Nuovo Cimento*. I had to conclude the job on behalf of SFP but the hard preparatory work had been done by my predecessors. The EPS is responsible for the editorial aspects but the publishing aspects remain in the hands of the French and Italian Physical Societies and there are many partners in the financial venture to share the dividends. Yet it works, it is a success, but still the only one. On the other hand, attempts to bring some cooperation among physics publications in central and eastern Europe, which otherwise run the risk to no longer find interest with hard pressed library budgets as

they have lost their Academy subsidies, have so far met with no tangible response. Yet they cannot all survive individually in commercial terms. We should try harder.

I have been editor for particle physics for North-Holland for 18 years, first with *Physics Letters B* and then with *Physics Reports*. The EPS was never in the picture. The same also applies to the successful physics journals of Springer-Verlag.

On Bernardini's list, one also finds summer schools. EPS started in '91, its series of summer schools "for southern Europe", bringing together prestigious lecturers but also relying much on local expertise. The first one ('91), on molecular physics, was in Spain. The second one ('93), on laser physics, will be in Crete.

Also mentioned is the publication of a regular bulletin. It took some time for *Europhysic News* to become a much appreciated publication. This now works. Next came the fostering of the teaching of physics. Much remain to be done but it is at long last recognized as an important priority. Last on the list was to favour the exchange of students and that of young physicists. A big effort has been invested recently, and the EPS has now a "Student Mobility Scheme" which should much help universities benefiting from the programmes funded by the European Communities. Over 120 universities have already joined the scheme in which EPS puts the needed human effort and seed money.

The goals spelt out by Bernardini were high and bold. They are still much valid today. Some have been met as the EPS prepares to celebrate its 25th birthday. Others still call for much enthusiasm, dedication and hard work. Yet this still illustrates very well all that one could expect from a physical society.

In order to bring a personal touch, I now go through some memories of the time which I spent with the SFP and then with the EPS, through a period which altogether covers almost a decade.

6.6. Three Years With the SFP

When one day, Anatole Abragam asked me on the telephone to become president of the SFP, I did not know at all what it meant but I could but say "yes" to him. At that time, the mandate corresponded to a three year period during which one was vice-presidents then, president and then vice-president. The president was therefore, always second bodied by two vice-presidents and had one year to learn the tricks of the trade. I succeeded to Philippe Nozières and was succeeded by Pierre Aigrain. I very much enjoyed collaborating closely with them during these three years. I still remember the board meetings where we three were puffing our pipes. This was considered quite acceptable at that time! During my last visit to the SFP, I noticed a big "No Smoking" sign in the meeting room. Things have changed!

The SFP has not a very large membership. It was at the level of 3000, which is little for a country like France. It could easily be four times larger, and it has not grown since. There are however, strong differences within Europe. There are countries where belonging to a professional association is felt as a natural attitude, and where most physicists join their national society. This is the case in Germany, in the Netherlands, and in the United Kingdom. There are however, countries where manifesting openly one's own individualism is felt as part of the national culture and where few people find it proper to join a society and pay their fee. This is the case in France and Italy. There are even countries like Greece, where few physicists thinly spread themselves over several competing societies. It is fun to live with such a variety but it makes life difficult for some societies.

A small membership does not provide enough income to rent a place, pay a secretariat, and run a bulletin, which is the minimum, a society should do. The main source of income of the SFP is its Physics Show which it runs every 18 months. It is a show of scientific and high-tech instruments which attracts many exhibitors and visitors. Much money comes in connection with it. Much money goes out organizing the show but the remaining profits allow the society to live in a proper way. My year as president turned out to be a busy one, with one physics show and one general conference, as there is one every other year. There were also the four council meetings and the board meetings almost every month. They were also the prizes to award with a well-organized committee procedure. There are always matters which one considers as a personal challenge. I wanted to push two things. The first one was a closer relationship with the powerful German Physical Society, the DPG. I felt that the SFP had but to gain in learning how the DPG was working. One of the first things I did as president was to visit the DPG. A joint prize was started, the Gentner-Kastler prize. It is now awarded every year, alternatively, to a French physicist chosen by the DPG and to a German physicist chosen by the SFP. The SFP had already such a prize with the Institute of Physics of UK, the Holweck prize, and the IOP had also one with the DPG, the Max Born Medal. The IOP is the physical society of Britain. Such prizes are very useful. They force people to better know their colleagues in other branches of physics in another country, when one knows mostly and almost exclusively, colleagues in one's own field. They are also the occasions of instructive meetings between officials of two different societies. The Gentner-Kastler prize was the first one ever with an amount labelled in Ecus. We also had a joint Franco-German meeting on *The Quark Structure of Matter*, which I could easily organize since it was in my own field. I did that together with H. Meyer who was the president of the High Energy Physics section of the DPG. We started in Strasbourg for 3 days. Buses then collected participants on the Sunday to take them on a tour in the Alsace vineyard and left them in Karlsruhe in the evening, for another two days of meeting. Few people thought that it would work but everything went very well. There were about 150 people attending the meeting. We got some funding, equally from the French CNRS-CEA and the German BMFT, and respectively handled by the SFP and the DPG. We spoke English and had the proceedings printed in Singapore, thanks to my friends at WSPC (World Scientific Publishing Company). It was a great experience. Such a "boy scout" spirit is good from time to time.

The other venture was to invite every other meetings of the board, a director of either the CNRS or the CEA, the two big research organizations in France. They all came with pleasure and apparently enjoyed an evening discussing the organization of research outside their usual hierarchy. It is amazing how misunderstanding can easily originate from limited information and how information has difficulty to travel from one organization to another. I think that we fulfilled a good role at channeling information and at listening.

During my year as president, the SFP organized a wide discussion on the "new thesis". This was a time when the government had decided that the thesis should better be reduced to two years. The pressure was apparently coming from Industry, which was said to be willing to recruit green PhD's provided that, they would not have gone through too long an academic contamination period. There was of course, an uproar in some academic circles. We wanted to assess the situation and express a SFP view. With the new scheme, anyone remained free to spend as much time as he or she would wish in completing his or her PhD. However, the thesis work is expected to be supported by a fellowship, which in Europe comes from a government source. If the government says that the fellowship should be limited to two years, it bears weight. Academia has to comply. I naively thought that opinions would differ according to specific fields of research, "heavy" science being for instance, in favour of a long PhD, and "light" science more open to a short one. But our survey showed that people were equally split for and against the short thesis, and this across all fields. The separation was actually found to be according to two different and equally valid points of view. Some felt that a thesis should be the proof that a person can continue doing research on his or her own, and can perform well as an assistant professor. A thesis of typically four to five years, as it was previously the case, was then considered proper. Others felt however, that a thesis work should first of all prove whether a person was gifted for research and worth being encouraged to continue. In that case, two years were plenty enough to judge. I had to report our finding to the Minister, advocating for some flexibility in time, whether the person was looking for a "training for research", and *a priori* willing to continue, or for a "training through research", with a career in Industry in mind. This is the way things eventually went. In '86, a change in government almost brought back the former thesis but the much wider scope bill was eventually dropped in front of students' demonstrations which led the minister to resign. At present, one sees indeed a convergence appearing between the different European academic systems with four years of undergraduate studies (Continental style) and two to three years only for the PhD (British style).

During my three years with the SFP, I met the Minister for Research four times and there were three of them. The political life in France at that time was indeed such that the ministerial turn overs were more rapid than those of the SFP! I learned what to advocate with strength at each time, three points:

- Do not focus on the present research and academic complements but maintain a decent influx of young people.
- Avoid important fluctuations in the yearly research budget. A big increase is even pointless if it is (likely) to be followed by a drastic cut.
- Avoid too strong a dichotomy between the so-called pure and applied research sectors. The meaningful distinction should be between excellent and "not so good" research.

In '84, the SFP president and vice-presidents met with Monsieur Fabius for a rather long meeting. He was very kind. We were all "Normaliens" weren't we? He hardly said a word. However, in the paragraphs on research of his introduction speech as Prime Minister, a month later, we found most of what we had said to him. We were certainly not the only ones to have influenced him. His Research Minister was Monsieur Curien who clearly shared the same views as a physicist himself. Nevertheless, we were very happy. Despite its relatively small membership size, the SFP can have some impact. French physicists who tend to ignore it should know that better.

6.7. The Two Years as President of the EPS

The SFP brought me to the Council of EPS. I was elected on the Executive Committee where I served for four years before becoming president in '91. It was wonderful at EPS to meet colleagues from many different countries, and to discover all that we have in common and all that we can do together. This is also very much, the case at CERN but the close knit community of particle physicists is rather special. Yet CERN is an invaluable microcosm in which to nurse ideas that may be quickly generalized to the whole of physics, even if there is often much inertia or lack of faith to overcome. One meets with some frustration but there are also great moments.

I met Albanian physicists and, in '92, we admitted the newly created physical society of Albania. I was surprised to see that the first two Albanian colleagues whom I met through CERN, spoke quite good English. When I inquired where they could practice it, they told me "with the Chinese". Contacts with China were for a long time their only allowed ones. Neither the Chinese nor the Albanian could communicate through their own languages, so they naturally used the one of their common official "arch-enemy". I met with Estonian physicists coming to Uppsala, where we had our EPS council in '90, to seek admission of their newly created society into EPS. It looked like the revenge of Charles the XIIth! They were also admitted in '92, together with the Lithuanians and the Croatians. I was in Budapest for the 100th anniversary of the Eötvös Society, which is the Hungarian Physical Society. Hungarian physicists had come from all over the world and, among them, Edward Teller. But there were also many neighbours, Hungarian speaking physicists from Slovakia, Romania and Karpathian Ukraine, all bringing their wishes for many happy returns, in Hungarian. There was the joint APS-EPS meeting in Budapest where we had representatives from all societies in central and eastern Europe. There were many great moments at EPS.

The EPS has a dual structure. On the one hand, it is a society of societies, 33 of them now. It is thus a federation. On the other hand, it is a fully international society through its speciality divisions and its committees where physicists work together as members of a unique European society. The EPS has also an individual membership but it is not very strong, 4000 in total. There is a tricky equilibrium to live with, with these various components, but this illustrates how Europe in its slow making process combines national institutions together with budding international ones, which only gradually increase their role.

Ever since it was created 25 years ago, the EPS tried to ignore, as much as it was at all possible, the imposing barrier which long stood across the continent. From its very beginning, it tried to bring together physicists from the western and eastern parts of Europe. This was not easy but some precious links could be established that way. It was indeed long, only through EPS that physicists from western and eastern Germany could keep some regular contacts. In '66, when one first talk about EPS, this actually looked ambitious but possible. This was a time of relative thaw and more and more conferences were bringing physicists from the east and west together. The number of exchange visitors was increasing. CERN had established good relations with Dubna. In '68, when the EPS is eventually set up, the abrupt end brought to the "Prague's spring" experiment which strongly damaged the collaboration spirit but it is clear that one should try to continue no matter what. This bridge between the two pieces of Europe has long been a very special feature of EPS. Now that the barrier is at long last gone, one should capitalize on the links established through difficult times in order to help collaborations bloom. Developing east-west relations as well as possible was, and still is, a big challenge and a clear priority for EPS.

An "East-West" task force was set up. In '92, it was then transformed into the "East-West Coordination Committee" to stress its importance and its perennial nature. The idea was to bring together representatives from the central and eastern Europe physical societies and find out with them the best ways to help. They wish to restructure their system, usually based upon an overwhelming academy, with an unhealthy separation between research and teaching. They wish to strongly develop their collaborations with the western part of the world. They need modern communication systems. They want their government to know that they did good things in physics even though, means were scare and that they can continue to do so. They need many things.... In '91 we had a good meeting in Dagstuhl, in the Saarland. It brought together representatives from practically all the major research organizations in western Europe and representative from physical societies in central and eastern Europe. The idea was to see whether there could be a western "model" worth following for the organization of research and also to explore the best ways to channel help with hard pressed budgets in mind. It worked very well. Each western representative described his own system with a mixture of pride and despair. Much pragmatism is required. This is in Dagstuhl that the "bottom-up" policy was considered as the only proper one. Help should be channelled "at the bottom", through collaborations between western and eastern groups organized along specific lines of research. The western research organizations were then agreed to grant extra supports in such cases. It was however, considered pointless to provide help directly through the existing eastern research structure, thus from the "top". There was a total lack of trust. The next step was to help people getting to know one another better. A directory of physics institutes in central and eastern Europe has since been published and another one for the former Soviet Union is being completed. One should thus make better known the facilities in the east which are now open to western users in order to avoid a one way brain drain. From mid '91 onwards, the problems brought by the situation of physics in the former Soviet Union quickly became overwhelming and extended much of our work first focused on central and eastern Europe. A good spin off of Dagstuhl was that the representatives of the western organizations found that it was very interesting to meet together. They have now set up a series of meetings for that purpose, where they invite an EPS observer. This is EUPRO.

During '91, strong, and I hope, long lasting links were established with the American Physical Society. This is particularly the case for all "east-west" matters. The eastern component of EPS was long a hindrance for too close relations with APS or merely an excuse not to develop them. But, when the then Soviet Physical Society and the APS started to collaborate, there was no case to leave EPS behind and links developed very quickly, thanks to the goodwill met on the American side. Through '92, this was heralded in particular, in three joint presidential letters published simultaneously in *Europhysics News* and in *APS News*. One of them is on coordination and collaboration on "east-west" matters. There was indeed the very successful "summit" meeting in Budapest in May '92 and in October, the EPS-EWCC and the APS-CISA met together in Amsterdam.

Helping our colleagues in CE-Europe and in the FSU remains a strong priority for EPS. So many difficulties have to be overcome to meet the present expectations and hopes. We initiated several actions, the major one being the help with journals and in particular, the "journals for Russia" programme.*
Next on the list is help with modern communication, meaning finding where stumbling blocks are and removing them. Every thing is now well coordinated with the APS actions and, whenever appropriate, we try to act together. We have also developed links with the Physical Society of Japan on those matters.

The new turn to be given to "east-west" actions and the new turn to be given to the APS-EPS relations and to the relations of EPS with the Physical Society of Japan and the Japan Society for applied Physics were great and interesting challenges during these two years. It was a beautiful team effort with lots of goodwill and enthusiasm met all along.

Physics is universal. Many physics ventures, conferences, of course, but also very large facilities have increasingly to be considered and some time organized at the world level. We should rely on our very good contacts with APS, the

---

*It is financed by INTAS, with typically 30 European Journals being delivered to Institutes in the former Soviet Union in '94 and '95.

Association of Asian Pacific Physical Societies and the Japanese societies, to do our best to foster such a worldwide cooperation and collaboration. Direct links now exist between the APS and EPS divisions and committees. A direct link exists between our ACAPPI (Action Committee for Applied Physics and Physics in Industry) and the Japan Society for Applied physics. It is hoped that joint workshops will eventually take place in increasing numbers.

A more down to earth challenge had to do with finances. The EPS budget was showing an increasing deficit, the so-called "red shift", and matters were brought to a head in '90. The society was kept afloat borrowing regularly, and more and more from some of its divisions which had some accumulated financial cushion of their own from and for their running conferences. This could not continue.

Membership fees paid by the National Societies are low per member, membership fees paid by individual members are not very high, and they have to be collected in various currencies, convertible but at a cost. It is always hard to negotiate a rise of the fees and it merely compensates for past inflation. Fees from eastern Europe were long paid dominantly in nonconvertible currencies and just "melted away" through inflation on bank accounts, since we had no ways to use them freely on the spot to buy western standard goods and even pay for hotel bills. We did buy a lot of stamps however, and did a good fraction of our mailing from Hungary, but not enough to absorb what should have been about 8% of our budget. Now, currencies can be converted but societies, short of an Academy subsidy, are too poor to fully pay their fees. It is well-known that the usual big source of income for a big society is publishing. However, in Europe, EPS came too late and in too old a world, where most of the physics publishing business is in the hands of powerful commercial companies who knew better than others how to adapt to a changing world, a few decades ago.

We transferred half of our secretariat activities to Budapest. This was the proper time to try the experiment and it also had a psychological effect. It worked, thanks to much goodwill and this brought some savings in running costs. However, it could not be considered as the proper long-range procedure. The main secretariat office has to remain close to the dominant part of the membership, with a barycentre clearly in western Europe, and this is now again the case. The key to a good financial situation seemed to be with Associates, namely, companies, organizations, and large laboratories with a strong activity or interest in physics. They each contribute a sizeable membership fee to the society. A big drive toward Associates was set up. It worked despite the grim economical situation which was leading some companies to drop their past support. We do hope that our Associates find with EPS a working partner with whom to adress many topical issues, and not merely an institution worth some help. In '93 we have a balanced budget, our accumulated deficit is gone, and we can provide seed money to specific activities. At the same time, *Europhysics News* is recognized as a very good magazine even though, it is run on a "shoe string" budget.

Another interesting challenge was to develop good relations with the European Science Foundation and with the Commissions of the European Communities, and in particular, DG XII which handles scientific research matters. I already mentioned the European Research Conferences. We met with much sympathy and interest. It is our hope that the EPS will soon take a stronger role in science prospective studies at the European level, something which was explicitly implied in the missions spelled out by Bernardini, but which had been set dormant for too long, emphasis being put on conferences. EPS has the competence and the legitimacy to rally constituencies to discuss large new projects at the European level. In '92 we had a very good meeting with our associates on the theme of "large facilities". We were guests of the ILL (high flux reactor) and the ESRF (synchrotron light) in Grenoble.

I also already mentioned what was initiated on education. Starting in '94, EPS will award, jointly with the Amaldi Foundation, the Amaldi prize for a physics text book at the full European level.

Edoardo Amaldi devoted much time to writing physics text books despite his many research and international activities. His vision, enthusiasm and drive were of utmost importance in the birth of several European research organizations. When arguing many years ago for a European space programme, he once said, "Nothing can best provide a link between a Sicilian peasant and a Norwegian fisherman than to know that there are in the sky, complicated objects which circulate around us and which in some way, belong to both of them." The same applies to quarks and leptons, doesn't it? Our hope is that the grandchildren of both, along with many other young people, will awaken to the beauty and challenge of physics, benefiting from the best talents borrowed from our many cultures.

a. "LOOKING FORWARD". ADDRESS OF THE PRESIDENT — 1991

The European Physical Society (EPS) is rather special. There are physical societies in all the European countries. The EPS is here as a framework within which these societies can act together whenever appropriate. It was created in 1968. This was a time when scientific exchanges between the western and eastern parts of Europe had been increasing and when one could hope that they would quickly develop. This was also a time when the success of CERN had already demonstrated what physicists from different countries could do together. There was a widely felt need to extend this collaboration spirit to other domains of physics and time was considered ripe to do something which would try to ignore, as much as it could be possible, the imposing barrier that was still extending across the continent. Whereas long and difficult preparatory work seemed to have make things possible, the invasion of Czechoslovakia, which occurred just before the society was to be set up in a formal way, almost jeopardized the whole operation. Thanks to much goodwill, the Society could still be created as foreseen and as scheduled.

The EPS has member societies and all physical societies in Europe are now associated with it. It has also individual (ordinary) members (IOM). They often join with the idealistic feeling that physicists throughout Europe can already act together independently of existing boundaries. It is true to some extent but boundaries are still very much present. It is sad that their number never became very large. It leveled at about four thousands.

Each Member Society is contributing to the budget of the Society according to a nonlinear scale which takes into account the number of its members. The fraction of the "unit fee", paid for each registered member, goes down by successive steps with the number of members. The representation to the

Council goes with the contributions, but again according to a nonlinear scale. The two very large societies, namely, the German and the British ones, do not therefore, take an overwhelming role as compared to the smaller, and sometimes much smaller ones. The Soviet Union had no physical society. It joined through the Soviet Academy with a rather limited number of declared members as compared to its actual strength in physics. It therefore did not take any leading role in the Society, but its presence allowed the other societies in the eastern block to fully play their part. Whereas the need for the EPS has never been challenged, in particular, for the organization of conferences at the European level, something which it does successfully through its many Speciality Divisions, a strong notion of subsidiary has to be kept in mind. One has to follow a narrow path between what has to be international and what has to remain national. Financial realities and problems quickly bring this back to the mind of whoever has a tendency to forget it.

In some of the smaller countries, many physicists would be willing to merge their society into a larger EPS. In the larger countries however, one most often meet with a strong willingness to keep the national society very much alive and there is no strong desire to merge the successful activities of the society into wider scale EPS ones. Since the larger societies are the larger contributors, one has to live with that. The EPS is to a large extent kept alive by their contributions but it is almost on a starvation diet. This is the more so embarassing that any society would like to launch new ventures of its own for which some financing is needed in the first place.

The financial situation of EPS has long been a very critical matter. The amount of red ink used in formulating its budget had a tendency to increase. This was even often referred to as the "red shift". In 1970, this had become really critical, with a large accumulated deficit which amounted to over a fifth of the yearly budget. This is in that context, which still prevailed in 1971, that the present address was given.

As a new president, I had the "priviledge" to try myself at answering the cardinal question of defining the role and usefulness of EPS, something which my predecessors had also previously to face with their heartfelt arguments. This address includes what I could say along that line while also stressing some directions in which I was willing to invest some particular effort and for which I wished to obtain the encouragements of Council.

Another question had started to become topical at that time. "Restructuring" was then considered as a good word. It has not yet been associated with the increase in unemployment usually linked later with its implementation in industry. The Society had to find its own way to "restructure", if only to face

106

its financial crisis. The idea was to associate a much larger members of the national societies to the activities of the European society. The members of its Speciality Divisions and of its specialized Committees were drawn from the individual (ordinary) members only, which represented only 8% of the total member strength of the member societies. Such activities could be open to all of them while the payment of a unit fee would also be gradually extended to all, thus departing from the decreasing scale according to which small societies were contributing far more per member than the larger ones. The smaller ones were complaining about it. The implementation of this restructuring scheme was to take much effort and to call for many discussions during my two years of presidency. It was not fully over in 1993 but it is now eventually coming into effect.

The EPS is a very interesting microcosm in the sense, so well illustrated by "The Masters" of C. P. Snow. One finds there, all the tensions but also the desires to achieve something together, which one finds among big and small partners in international and national life. The yearly budget is modest. It is of the order of $800 000. But arguments are as fierce as those met in much larger organizations dealing with overwhelmingly larger budgets. Things may be easier with physicists since they share a common culture and are known to have a rational mind. Yet what is happening within EPS may be interesting on a wider scale.

A Statement by Maurice Jacob on Being Elected President of EPS

I cannot grow
I have no shadow
To run away from.
I only play   W. H. Anden, Hymn to St. Cecilia (Britten)

We have problems but we can grow. However, the key question which we so often hear is: "Why should I join the EPS?".

This question actually covers two deeper questions, namely, "What would the EPS do for me?", and "What is the EPS actually doing?" — the latter probably being the more important of the two. The EPS does not yet have a visibility we would like to see associated with a thriving Society. Within the physics community at large, its name does not always immediately bring to mind proposals, activities and specific actions which clearly bear witness to its presence on the European scene.

I feel it my duty to present some still complicated answers to these simple questions. We, of course, all have some answers. The problem is that, despite them, the questions remain very much present.

I appreciated the positive vote in Council in favour of raising the unit fee. It recognizes the efforts made since the 1990 Council, and offers encouragement to continue the work. Council has also strongly expressed its wish for some important restructuring. The goal is clear. All physicists in Europe should find it natural to be affiliated with the EPS. Ways to achieve this have still to be found. The question should be addressed with a renewed energy. As clearly summarized during Council, our Society should be strong in the eyes of physicists. It should also be strong outside physics, and recognized as the natural partner by national and international organizations when considering physics in Europe.

EPS on the European Scene

Hang together

Becoming a member of a physical society reflects a wish to be associated in a common endeavor bearing directly on our scientific interests. Membership also often translates into a willingness to become more personally involved in activities upon which physics thrives. Physicists do not work in isolation: their research implies a network of conferences, workshops and committees in addition to extensive discussions among themselves and with funding agencies. Whilst they often work in specialized areas having their own styles, structures and organizations, most are deeply sensitive to the great unity of physics. More than a profession, physics represents a way to look at the world and address profound issues using approaches extending beyond any obvious subfield and any national or cultural boundaries.

There is clearly a need for a broad approach to all questions associated with the development of physics as it forms a part of human culture and is a key element of our technological society. It should most clearly be met by a physical society. We all know that research has to progress on a wide front, and that harmony should prevail between the different domains of physics. As Benjamin Franklin once said: "Let us hang together or we shall hang separately".

Defining a role

The need for physicists to organize themselves, to work together, and to speak with one voice, is certainly at the origin of the great success of some national societies in Europe, notably those whose membership includes most of their country's physicists. However, in other countries, a more individualistic attitude still seems to prevail, and society memberships include but a limited fraction of the total number of physicists. This does not mean that the majority of physicists are satisfied with a relative isolation. Physical societies are not the only structures available. There also exist important authorities which often act as nation-wide organizations for research, speaking for physics at a government level, allocating funding and responding to specific initiatives. The development of these highly structured organizations, which complement the universities, arose from the growing importance of physics research. The increasing internationalisation of physics research also led to many links between national agencies and to the emergence of international organizations.

Being associated with one of these organizations may be considered enough, and all the more so as this does not imply a fee. Physical societies have had to define their role with respect to them. They do this with varying success.

There is a need for channels, running in parallel with such important and efficient structures, which respond to the grass-root support for new initiatives, and to questions and problems that are felt deep inside the physics community. These channels should operate independently of those offered by the official organizations, with their hierarchies and specific interests. Physicists should be able to express themselves and organize their collaboration independently of the often specialized and hierarchical structures to which they often also belong. National societies respond to that need, but with very different membership successes.

Pan-European

While physicists have always been keen to cooperate across national boundaries, they are now forced more and more by the evolution of their technical fields to conduct research in an international way. In areas where this is already well established, they have long realized how fruitful it can be. There should therefore be a feeling spreading among physicists in Europe that they belong to a European-sized community which expresses itself directly through EPS.

Being associated to a European-sized organization is certainly part of the motivation to become an IOM. There is even a "European dream" aspect, that is, a willingness to use physics to build a Europe which would eventually extend beyond the present achievements of the European Community, or even the wider circle of the some European research organizations.

Professional appeal

We have to acknowledge, unfortunately, that the number of IOM's is still relatively small. The European dimension of our community is not yet widely perceived at the individual level. There is an interesting point to notice in this context. In countries where national societies attract most of the active physicists, the motivation to also join the EPS is not yet present on a wide scale. On the other hand, in countries where national societies recruit a relatively small fraction of the physicists, their members are far more likely to also become IOM's. There is, for them, in joining their national society a strong vocational component, besides the purely professional one. It then easily extends to the EPS. The EPS should better respond to the hopes of those who join it with a European vocation by also becoming an invaluable professional tool. It should gain a visibility such that it is also clearly appeals to those who join a national society to first manifest their professional affiliation.

Complementarity

There is much to be done. But there is hope for success. Here comes another fact. In domains where European collaboration is already almost complete, the EPS is also very much present. Indeed, in view of structures for cooperation such as CERN and ECFA, one may wonder whether there is still room for the EPS! But one remarks that it is those physicists who apparently need the EPS the least who in fact use it to the full.

The reason may be that they already have experience in collaborating at the European level so they can use efficiently the extra and special possibilities offered by EPS. As an example, the very successful European High Energy Conference series started independently of the EPS back in 1961. Yet it eventually became natural that it should continue within the EPS, as it has done for many years. Similarly, CERN provides a very efficient structure and yet the HEPP (High Energy and Particle Physics) Division is particularly active.

EPS thus complements existing European structures. In other domains, where European collaboration is far less advanced, the Society already provides some of the needed links. It has made its mark in running successful conferences and providing forums for discussion. This is especially important in areas where the use of international facilities, each serving a large number of users, is set to increase. Decisions to proceed with new facilities will rely more and more on the existence of a clear constituency of potential users. The EPS undoubtedly has a great role to play in helping them manifest themselves. It can provide the framework and legitimacy for the organization of meetings where discussions and assessments should first take place.

EPS has thus a important part to play through its speciality Divisions, which represent the key component of its activities, and its committees, whose work allows the Society to speak for physics and physicists at the full European level. Success also has much to do with the national societies, which represent the other component of the Society's dual structure that makes EPS quite different from a mere federation of national societies. EPS should act in close association with its Member Societies, which represent the largest fraction of its present grass-root support.

Physical Societies — Past and Present

Traditional benefits

In the past, a physical society was basically a learned society where members met to exchange research results and information. Each society was an island of specialized knowledge in a world which considered science as a particular intellectual exercise. As they grew, societies started publishing journals and running conferences. Membership meant recognition by one's peers, advantageous subscription to specialized journals and the access with a reduced fee to conferences, with, in some cases, the right to present a communication. Societies manifest their specific presence by a bulletin, award prizes, act as forums for topical discussions, make their views known and put them to action by initiating new ventures, either at the advisory level or at a more concrete and direct one.

Things of the past

Whilst initially organized on regional or national levels, the different physical societies adapted with different successes to a rapidly changing situation. In Europe, the post-war period has been a particularly challenging.

Peer recognition and assessment take place through universities and research organizations far more than through physical societies. Physics has largely become a business. In Europe, commercial publishers adapted more quickly and efficiently than most societies to the changing scene of having to operate journals on an international basis. The prevailing culture, based upon cheap photocopying, high subscription prices and the desire to protect one's personal shelves from an overwhelming invasion, is such that subscriptions to scientific journals have practically become a library business. An inexpensive subscription is no longer an incentive to join a society.

The same applies to conference fees. There are so many conferences that one attends only those from which one is likely to benefit most. Once this drastic choice has been made, support from laboratories has become rather widespread. So reduced conference fees are no longer a strong incentive to join a society, and the more so now that there are many important conferences run outside the control of any physical society. The big general conferences, which have still remained more closely associated with physical societies, are increasingly shunned in favour of "stand alone" conferences and topical meetings or even workshops. It is clear that reduced subscription rates and conference fees, long seen as incentives to join a physical society, will soon belong to the past.

Opportunities

The very same reasons which push these advantages into obsolescence, provide, however, more strength to those incentives associated with being part of a society which has a recognized advisory role, and plays an active part in matters relevant to the life and evolution of research and of advanced teaching in physics. It is therefore in these directions that EPS should develop its activities. Our Divisions and Interdivisional Groups should respond even more quickly and efficiently to precise needs in their fields, so that their actions can be clearly perceived within their research communities. Our Action Committees should expand their activities as forums for discussion and as the originators of specific advice and actions, so that the European physics community at large sees there is a powerful way to make itself heard.

A modern physical society thus has three important roles to play:

— It should contribute to the development of research though its advisory role with funding agencies and through the organization of conferences and meetings.
— It should foster a better synergy between pure research and industry, for a sharp dichotomy between pure and applied research has many disadvantages.
— It should help in making choices for the formation of physicists and for a better appreciation of physics research by the public at large.

The main advantage the EPS can offer its members must be the possibility to play an active part in a thriving hub of activities which meets these three requirements at the European level. The Society should provide a frequently used base for those interested in launching new activities. Some of our recent initiatives, reported in *Europhysics News*, stand as useful examples.

112

My past experience with the French Physical Society gives me confidence that a society, despite, but rather because of, its lack of executive power, is listened to. Deciders like to hear from an independent forum, and they also appreciate finding a sounding board for new proposals which responds independently of any of the existing and powerful hierarchies through which they manage research.

The life of the Society has to also find direct and easily accessible expression through its bulletin, which should be deemed worth reading. *Europhysics News* has improved, in particular in its coverage of topical questions. It will be possible to develop further, even in the present framework of adiabatic expansion.

The Society met success with *Europhysics Letters*. However, the publishing market in Europe is largely in the hands of efficient commercial publishers and as we lack capital, there is little that one can do at present on the purely publishing side. This is sad since this is a most natural source of income for physical societies. Given the present trend to "farm out", it is perhaps not even clear that we should try to expand in this area. Nevertheless, much more can be done in participating at the editorial level in existing and future physics journals.

Where to Act

The EPS is a society of societies as well as a society of **members**. Seen from the individual's point of view, this is often not very satisfactory and the question of the two separate memberships which it implies has been repeatedly addressed. Questions also repeatedly come from both components of our dual structure. Some national societies would like to consider the EPS as a federation of societies, with membership of a national society coming first and implying direct membership in EPS.

However, some Divisions which organize very successful EPS conferences and workshops are annoyed to see that only a few participants are IOM's. Most participants appear satisfied with membership in a national society, or, actually, with no membership at all. Special fees have, as we saw, a declining incentive value: there is no point playing too much on them. The Division Boards would like to see only one kind of membership, but which would then be an EPS one to start with, since it would be linked to a specialized Division activity in the first place. The hope is that a sizable increase in EPS membership could then snowball. From this point of view, national societies would then appear as "national groups" within the wider structure provided by EPS.

The question is difficult since the unique membership solution appears quite different when approached from one or the other of the two components of our dual structure. But they both raise it, so the question has definitely to be addressed. There is clearly a mandate from Council so it will be the main item on the agenda of the Executive Committee. It is our hope to approach the national societies and the Divisions with proposals in time for a discussion at Council, in Athens, next March.

The EPS should also provide closer **links** with physics in the world at large. The Society should, in particular, have stronger specific and privileged links with the American Physical Society and with the newly created Association of Asian Pacific Physical Societies. These links, especially with the former, would become more lively if they directly involved the Divisions. The aim then would be to better coordinate the organization of conferences and to run topical joint workshops. Much remains to be done. The interest expressed by Council was clear. It is our hope to implement initiatives as quickly as possible. Associations with other large physical societies in the world, who consider the EPS as their natural partner in Europe, should be a further incentive to join the EPS once their outcome has become lively and visible to all.

We should develop our links with our **Associate Members** into an active partnership. They represent a most valuable connection with applied research and industry. They certainly represent part of the solution to our financial problems.

As Benjamin Franklin wrote in *The Way of Wealth*, reprinted 400 times: "There is no gains without pains. If we are industrious we shall never starve; at the working man's house, hunger looks in but does not enter, nor will the bailiff or the constable enter, for industry pays debts while despair increases them". Let us work together for a brilliant 25th Anniversary of EPS in 1993.

Published in Europhysics News, Vol. 22, pp. 107–109,
© 1991 Bulletin of the European Physical Society.

b. ADDRESS TO THE COUNCIL OF EPS — 1992

The 1992 meeting of the Council took place in Athens. The restructuring questions, which had come up in 1991 were an important part of the agenda. No decision was foreseen at that time, since many financial problems had first to be solved. Nevertheless, the option considered was much under debate and it was concluded that the direction was correct. The discussion focused also on East-West matters. The societies had taken several steps to help physics in central and eastern Europe but help to physics in the former Soviet Union was quickly becoming the major problem. Good contacts had been established with the American Physical Society and soon afterwards, there was actually a "summit" meeting in Budapest, attended in particular by the presidents and vice-presidents of the two societies. Under discussion at that meeting were all questions on which the two societies could act together in their efforts at helping colleagues in the eastern part of Europe fulfilling their hopes but, first of all, surmounting their present difficulties and predicaments.

I had long felt that EPS should establish a good relation and, eventually, come to some sort of partnership with the Commission of the European Communities in Brussels and also with the European Science Foundation in Strasbourg. Progress along these lines were reported. This address shows how it looked in 1992, before the Council meeting in Athens.

The President's Report

"Drive thy business, let not that drive thee"
Lost time is never found again"

Benjamin Franklin

These two quotes suit well the Society's present situation. We have already given some thought to a restructuring which, following the wish expressed by Council last May, will be thoroughly discussed at Council in Athens next month. While it is clear that we operate under very strict financial boundary conditions, we ought at least initially know what we *should* do, rather than merely reflect on what can be done with what we presently have. For we should try to drive the Society and not be driven by its presently limited resources. This is one thing. But the most important aspect at present is the proper exploitation of the golden opportunities offered by the disappearance of barriers which divided Europe for so long. Expectations are great: the difficulties speak for themselves.

Fostering east-west collaboration remains the main and urgent task of our Society. There is no time to be lost. We have to continue identifying the best ways for action, and then act: we should feel privileged to live during such a challenging but promising period.

While these themes are preoccupying, I shall turn first to more routine matters. EPS has maintained a good level of activity throughout 1991 — activities that were to a large extent those of the Divisions, Interdivisional Groups, Action Committees, and task forces, each operating at the full European level. Many conferences, meetings and workshops were organised, detailed reports of which will be summarised in *Europhysics News* after Council.

In focusing on a few topics which concern the Society as a whole, I note that I have already served for about a third of my anticipated mandate, so this is the proper time to present the lines of action which have been initiated, and to indicate how their continuation is perceived at present. While some balls have been set rolling recently, many of the actions taken in fact follow initiatives started before May 1991 when the last Council took place.

Restructuring

The Executive Committee has been working on restructuring following the mandate of Council. We were given the task to propose a suitable scheme, whereby all individual members of the Member Societies would also become directly affiliated with the EPS. The main purpose of the restructuring should be to increase the visibility of EPS, both within and outside the physics community. A Society of some 60 000 members could play a strong role in the life of physics in Europe. Divisions and Committees operating from a base ten times larger than the present one would be able to increase their activities considerably.

116

So restructuring was a very important item on the agenda of the Executive Committee meetings in Geneva, in August 1991 and again in January 1992, and in Budapest last October. Much discussion took place in preparing for them and a position paper drafted by the Executive Committee was sent at the end of 1991 to the Presidents of the Member Societies, to Division and Action Committee Chairmen, and to the IOM delegates. In proposing a restructuring scheme, it remains first of all a *consultative* document. Benefiting from reactions, the Executive Committee will write a working paper as the basis for discussion at the next Council, where it will be an important item on the agenda. If, as we may now hope, a line of action can then be defined, we shall continue our work in order to prepare a scheme which could be proposed for approval at the following Council in March 1993, with implementation foreseen on 1 January 1994.

Many challenging questions still call for satisfactory solutions and the boundary conditions are tight. We should not rush but keep working. The working paper lists the lines of action along which the EPS could extend its activities as a result of its increased strength and visibility. The document proposes that, according to the new scheme, all members of the national societies would receive *Europhysics News* as a clear sign of their membership, and all would be eligible for election to the Boards of Divisions, Interdivisional Groups, Action Committees, and task forces. The proposal tries to minimise the extra financial support required from the National Societies while aiming to even out their contributions per member. A national society joining the scheme would become a "full membership society" and it is anticipated that all will eventually join, although the scheme allows for a gradual implementation. Representatives from the Divisions would be given a larger relative weight at Council.

After this brief survey of the present state of our work on restructuring, I shall very briefly turn to finances and to the operation of our Secretariat. This is limited to generalities: reports to Council by the Treasurer and the Secretary provide more detailed information.

Finances

Restructuring was partly motivated by our past financial difficulties. There are, however, limits to the support which we can receive from national societies in the present circumstances. The line followed is therefore indirect: we aim to achieve an improved financial situation from increased visibility — from a larger membership but most of all from a greater visibility in term of actions. The proposed scheme would help greatly if adopted in its general lines, but

117

we still need to work out a solution for what we call in physics a "bootstrap" problem to have the scheme move forward. Our finances are now in good order and we expect to have cleared our accumulated deficit by 1993. However, the situation still needs much attention, especially since the present operation of our Secretariat with the Executive Secretary based in Budapest should be considered as a temporary mode. The financial situation is regularly monitored by a finances task force having members of the Executive Committee as its permanent nucleus.

At present, some 36% of our resources come from the Member Societies, 24% from our Associates, 22% from IOM contributions, and 13% from *Europhysics News*. The finances task force has played a particularly important role in working to convince Associates of the importance of helping directly the Society's initiatives through financial and other forms of support. But help from everyone is very much needed! Keep trying to convince colleagues to become IOM's. Try to convince companies or research organizations to become Associates. The situation has already improved but we hope to make still more significant progress. *Europhysics News* assists in making what we are doing more widely known. Present increasing interest in our bulletin should be a great help in enticing others to join the Society.

An appeal for sponsors

We consider that all fees should now be paid in Swiss francs, or the equivalent, but we do not want to have a two-tier society with two or more levels of fees. We should take as a working hypothesis that the present great disparity in income between our western and eastern members will eventually decrease. Nevertheless, changes will be slow and we should show much understanding towards our colleagues from central and eastern Europe. I therefore turn to the richer societies in the west and ask them whether they would partly sponsor their partners in the east for a limited time so that fees are properly paid. Including the former Soviet Union, we are talking about 70 kSFR each year, out of which we shall probably have to find a significant fraction through sponsors during the coming two years. If, as I hope, our Member Societies are willing to help, this global figure indicates the amount which could pledged for a special entry on our budget. While the Society will cope with any shortfall, we should consider it a matter of principle that all fees are properly covered.

The situation is even more critical for IOM's since fees presently represent an excessively large fraction of a physicist's salary in central and eastern Europe. In this case, we are talking about 15% of our IOM's. I now turn to our IOM's in the west and ask them, "Are you willing to cover the fees, or part of the fees, for a colleague in central or eastern Europe for a year or longer?" An appeal along these lines has already been launched by Professor Buckel in Germany. Perhaps others would consider doing the same elsewhere?

Once again, collecting full fees is a matter of principle, quite independent of the resources they represent. As discussed later, the Society as a whole should find the best ways to help keep physics alive in the east. We are busy at it. A gesture from the Member Societies and the IOM's would clearly demonstrate the great concern for the present situation, and would be an extremely valuable encouragement to all.

Full involvement

We are ambitious when it comes to actions. But we have limited resources. Indeed, because of the present difficulties with research funding, many physicists are encouraged to take an early retirement. We should try to take a positive look at the problem by appreciating that there is a gold mine of competence and talents which we would like to exploit to the full in fostering actions by our committees and task forces. While meeting travel and ancillary expenses always remains a problem, it is nonetheless extremely valuable if we can call upon the time and goodwill of experienced physicists. To all those that this may concern, I would like to stress that our Society offers many opportunities for action in domains where it wishes to expand activities.

The Secretariat

Speaking in general terms, the operation of our Secretariat, with both a Geneva and a Budapest component, has been working as well as one could hope. Quoting again from Benjamin Franklin, I should say in this respect that "It is easier to build two chimneys than to keep one in fuel". Yet we managed. We should, however, acknowledge the difficulties which have to be met by our Executive Secretary and the rest of the staff. Their work and cooperation commands praise. Our goal is to bring the Executive Secretary back full-time to Geneva by the end of 1992. By then, the Budapest branch will be able to efficiently deal by itself with some specific tasks. At present, and for a few years, this branch represents an opportunity to quickly extend, if needed, the tasks entrusted to the Secretariat.

Some Specifics

I now turn to some specific points which are worth a special mention this year. In presenting some of our activities, the list is by no means exhaustive. More extensive and timely reports can be found in *Europhysics News*.

East-west collaboration

East-west collaboration has always been a focus for the Society's activities and will remain so in the foreseeable future. The recent political developments have brought wonderful opportunities and raised great expectations, but the difficulties are all too apparent. Help to colleagues from central and eastern Europe was mainly channelled through the Action Committee for Physics and Society (ACPS). Responding to their request, as formulated in the framework of our east-west task force, which has become a very efficient forum for the exchange of ideas and defining initiatives, the ACPS organised a workshop in Dagstuhl, Germany, last August. It brought together representatives of the main research organizations in western Europe and representatives from national societies elsewhere. It was a great success, being instrumental in defining the best ways to find financial support and to foster collaboration according to a "bottom-up" procedure. We are now trying to help along such lines as much as we can: a Directory of research institutes is being set up and assistance for international peer reviews will be implemented. ACPS has also monitored help for libraries and for instrumentation. Contacts with the American Physical Society (APS) have shown that the APS is eager to collaborate with us in helping improve the situation in central and eastern Europe and we are actively following up this possibility.

Visits to Poland, Czechoslovakia and Hungary have been the occasions of very interesting and constructive discussions. We organised the annual meeting with our Associate Members in Cracow, in September, in conjunction with the "Physics for Industry, Industry for Physics" Conference, while the meeting with the Chairmen of the Divisions and of the Action Committees took place in Budapest, in October.

Physical societies in Albania, Estonia and Lithuania have applied for membership. After hearing about their activities and reading their Constitutions, the Executive Committee decided at its October meeting to recommend them for membership to Council. Decisions should be reached in Athens.

The situation of physics research in the former Soviet Union is a matter of great concern. While we maintain our long association with the Soviet Academy of Sciences, we are also in contact with the Moscow Physical Society, the Soviet Physical Society and the recently formed Russian Academy of Sciences. We are exploring ways to help keep a vigorous physics community. Much the same can be said about the disturbing situation in Yugoslavia, where physicists are represented in EPS by the Union of Yugoslav Societies of Mathematicians, Physicists and Astronomers and by the Institute "Ruder Boskovic".

We have been staying in close touch with CERN on east-west affairs as the organization plays a leading role in the scientific integration of Europe. Our thanks go to the NATO Scientific Affairs Division and to CERN for the valuable help provided to our east-west task force.

North-south collaboration

Special mention must be made of the very successful 1st EPS Southern European School of Physics which was held in September in Avila, Spain. It benefited much from support from Spanish colleagues. The school dealt with topics in molecular physics and a software teaching aspect, organised with the help of IBM Madrid, was an interesting and valuable experiment which should be repeated. The next school, on laser physics, will be in Crete, Greece, in 1993. These Southern European Schools are somewhat special among the many summer schools as students must be fully supported and the topics covered have to promote research already well implemented in southern Europe. Funding comes mainly from UNESCO, the European Community (EC) and local sponsors.

Relations within Europe

We have remained in close contact with the DG-XII of the Commission of the EC. Approval of the new Framework Programme for science has been delayed, but we hope to be involved with its implementation in physics as soon as this becomes appropriate. Visits to Brussels are being maintained and *Europhysics News* regularly offers information on opportunities in physics. Our direct contact with Brussels is operating very well.

The outcome of visit to the European Science Foundation (ESF) in Strasbourg last July is summarised in a report which has been circulated. It was the occasion to itemise many points on which the EPS and the ESF could coordinate actions. The ESF is eager to work with us in physics. Special mention should be made of the organization of the European Research Conferences (ERC's) that are operated in the framework of the ESF with EC funding. EPS agreed at the beginning of the scheme to integrate its Study Conferences that had been running for 20 years. However, owing to some restrictions on organizers, it was decided in 1991 to continue holding separate Study Conferences. EPS continues to act as the adviser to the ESF for physics, and ERC's are proposed and run by our Divisions. We are happy that six ERC's in physics took place in 1991 and 10 are scheduled for 1992. It is hoped that such activities will greatly increase during the course of the next DG-XII programme when it is finally approved.

The progress of the EC Erasmus and Tempus mobility programmes are being closely monitored and one would hope to be increasingly involved with their implementation. We are working on student mobility and on physicist's qualifications at the European level. For the latter, fields close to health and safety require special and urgent attention.

EPS should appear as the natural partner when it comes to pan-European physics. Owing to limited resources, we must sometimes rely on goodwill and direct help. Nevertheless, we should try harder since this is the best way to increase visibility and eventually resources and means of action. Much should be done in close collaboration with Member Societies.

The EPS should also offer an effective forum for shaping constituencies for the most promising, new, international research facilities. Initiatives in two directions, nuclear physics and supercomputing, have been assisted in 1991. This is also a domain where we could efficiently act in association with the ESF where our help would be welcomed. The Divisions and Interdivision Groups could perhaps think about such matters and there will be a speedy response to requests and suggestions.

Relations outside Europe

As indicated previously, the APS is very interested in acting in association with EPS, particularly for all matters having to do with physics in central and eastern Europe. Two visits to the APS in 1991 were occasions to find ways to develop a fruitful collaboration. Following approval by Council in May, a scheme with link-persons between each of the APS and EPS speciality Divisions, and between each of the APS and EPS committees, has now been implemented. The aim is to develop a good flow of information directly at the action level. This should lead to improved coordination in organising topical conferences as well as be conducive to joint workshops and common actions.

EPS is still poorly known in North America, and yet one meets many colleagues who are interested in what we are doing when the information becomes available to them. The APS is helping us in our present drive for subscriptions to *Europhysics News*. The directions given to the collaboration between APS and EPS are treated in a joint paper by the Presidents of both societies which will be published simultaneously in the March issues of *Europhysics News* and *APS News*, the new bulletin of the APS.

We have also implemented a link system between the Executive Committee of EPS and the Board of The Physical Society of Japan, and are keeping in regular contact with the Association of Asian-Pacific Physical Societies.

New divisional structures

It is necessary to single out the former Optics Division and the existing Astronomy and Astrophysics Division for detailed discussion. The Optics Division decided to disappear as such in early 1991. This was not a surprise as a more industrial oriented, worldwide framework was perhaps better suited to the organization of the main conferences in optics. With the creation of the European Optical Society, the Quantum Electronics Division and the Atomic and Molecular Physics Division agreed to cater to the needs of physicists working in optics. The QE Division changed its name to the Quantum Electronics and Optics Division and all members of the defunct Optics Division have been notified. While what has happened illustrates the difficulty of having a lively structure where both academic and industrial research can satisfy common interests, physicists in basic research in optics should still be able to use the EPS for their needs, and this within our existing Divisions.

The recent creation of the European Astronomical Society (EAS) has been a serious challenge to our Astronomy and Astrophysics Division, the more so because its Chairman became a leading member of the EAS. In view of the creation of the EAS we have certainly to drop astronomy from our responsibilities. But we should by no means renounce astrophysics! The A. & A. Division does not wish to continue in its present form so we should develop an Astrophysics Division. Now that the EAS exists, we should do it with them rather than in competition. The idea is therefore to create a strong Astrophysics Division that should be associated with both the EPS and the EAS. Extensive discussions have taken place throughout 1991, in close association with the Solar Physics Section, and a report was presented to Council last May. The Section agrees fully with a draft proposal and it will assure the interim until the new Division is in place, encompassing, let us hope, several Sections covering most of astrophysics. Together with the President of the EAS, I have signed a proposal for the creation of a new Division that will be submitted for approval to the Councils of EAS and EPS. It seems that it is what best meets the wishes of astrophysicists in Europe, who, depending on their respective country, lean more toward physics or more toward astronomy. The Division would have members of either the EPS or the EAS, preferably both, and each of its Sections would choose either EPS or EAS for administration.

There remain "border" domains in which we should extend our activities, namely, mathematical physics, biophysics, *etc.* Suggestions on how to participate in these areas would be welcomed.

Publications

Publications are an important source of income for several physical societies. But this is one area which the EPS can hardly exploit as, in Europe, scientific publishing is to a large extent in the hands of powerful commercial scientific publishers. Moreover, we cannot compete with some of our Member Societies which are also active in publishing. We should instead help them reduce the number of journals.

Given the situation, which is not conducive to extra income, the EPS should increasingly apply itself to general topics in physics publication, such as electronic publishing, as is being done through the Publications Committee. The Society as a whole should also try to play a larger role on editorial boards as well as help coordinate the many different physics journals in central and eastern Europe which now seek a wider basis. These journals probably cannot develop without some specialization, which we could help define.

EPS-9 general conference

Preparation of EPS-9, with R. A. Ricci as the Conference Chairman and E. Brezin as the Chairman of the International Programme Committee, is now well under way. The conference, which will also mark the 25th anniversary of EPS, takes place in Florence on 14–17 September 1993. Falling between two other EPS conferences in Florence, one organised by the Quantum Electronics and Optics Division (11–13 September) and the other by the Liquids Section of the Condensed Matter Division (18–22 September), should represent a plus. One hopes the conference will bear witness to a thriving Society which has a high visibility.

Published in Europhysics News, Vol. 23, pp. 34–36,
© 1992 Bulletin of the European Physical Society.

c. ADDRESS TO THE COUNCIL OF EPS — 1993

The 1993 meeting of the Council took place in Nice. Restructuring was again a key issue. The discussion did not lead all the way to a decision on rapid implementation but good progress was made, with much goodwill and understanding on all sides. An important good piece of news was that the financial situation was at long last balanced and that EPS could now consider a balanced budget while launching at the same time, though in a still modest way, some operations of its own. An important element in reaching a sound financial base had been the support received from Associate Members. They correspond to industries, organizations and physics institute which each contribute a sizeable annual fee. Many of them could be convinced that the EPS was worth supporting and started to do so or increased their previous contributions. Others dropped out but that is a fair game. The contribution to the budget from Associate Members was brought to about 25% of the total. This helped a lot in reaching a sound financial situation. The role of the treasurer, Philippe Choquard, had been particularly instrumental and starting in 1990 already, the financial situation had been closely monitored by a specialized task force.

The involvement of EPS in education was shifting gear. This is a domain of important concern throughout the industrialized world in which some cooperation with the American Physical Society and the Physical Society of Japan was soon to start. This address shows how it looked in 1993, as I was soon to leave the inner circles of EPS.

All that may appear as but a modest amount. There is so much to do and everything takes time. Nevertheless, this address as the previous two, illustrate in a rather direct way how the EPS works. They may contribute to make better known the activities and, by the same token, the usefulness of the Society.

125

The President's Report

There is a tide in the affairs of men which, taken at the flood,
leads on to fortune.
Omitted, all the voyage of their life is bound in shallows and in miseries.

From William Shakespeare's *Julius Caesar*

This well-known quotation is far too strong to apply to our restructuring discussions, as most of the changes to be discussed at Council in Nice in March will in fact be adiabatic in character. It applies, however, very well to the solidarity which western physicists should manifest toward their colleagues in east and central Europe and in the former Soviet Union. The imposing barrier which long stood across our continent is gone. We live a time of great hopes and great expectations, but it is also a time when great difficulties have to be overcome.

Physics offers an invaluable microcosm in which to nurse ideas that may eventually help foster understanding and collaboration. For we speak a common language and share a common passion. Clearly, therefore, developing better links between the various parts of Europe remains a key priority for our Society. Despite the strong common bounds provided by physics, many cultures and languages keep us somewhat different — something which is both a hindrance and an asset.

The year may thus herald some changes in the life of our Society if the present restructuring scheme is approved. The hope is that this step will foster a stronger feeling of unity within the European physics community. We enter 1993 with a balanced budget and an accumulated deficit which is practically gone, a year ahead of schedule; one may thus look to the future with some ambition.

1993 is an important year for EPS as it marks the 25th anniversary of the Society. The occasion will be celebrated in Florence during the EPS-9 General Conference at which many distinguished colleagues, including four Nobel laureates, will give the plenary lectures. A series of major symposia covering in depth topical new developments will accompany the plenary talks.

Many discussions have taken place at many different levels since we started work on restructuring after the Zurich Council meeting in 1991. I do not think that what I say separately about the process goes against any majority or maybe consensus opinion, even though the emphasis put on some of the

126

themes reflects a personal point of view. The present report focuses instead on some recent, general activities of the Society, whereas those of the Divisions, Interdivisional Groups and Action Committees, that once again bear witness to a highly visible presence, will be covered in comprehensive reports to Council and summarised soon after in *Europhysics News*.

Finances

Thanks to information collected and commitments received since Athens, the restructuring scheme can now be proposed on a sound financial basis, along these lines presented in the consultative document. We should be on firm ground when the scheme is fully implemented, as well as during the transition period. A few words about finances are in order.

The Society's finances have been known for some time to be rather precarious and the 1990 Council meeting in Uppsala brought matters to a head. A major effort to restore the situation started the same year with the setting up of a finances task force which has continued working ever since. There was no magic wand to wave. Many thanks are due to the task force members, to our indefatigable Treasurer, Ph. Choquard, and to our Executive Secretary who transferred to Budapest for two years, together with a part of our secretarial activities.

Another important element in our reaching a sound financial situation has been the support we receive from Associate Members. While several industrial companies dropped their support owing to the present grim economic climate, many research institutions and large laboratories have agreed to join, and those Associates which agreed to stay have increased significantly their contributions. It has indeed been heartwarming to see many new Associates joining on the basis of what the Society is doing. I firmly believe that in Europe, where the physics publishing business is largely in the hands of powerful commercial companies, a lasting and increasing support from Associates is a key element of sound financial prospects. We also appreciate that there are limits to what we can ask from our societies and individual members. I would like to warmly thank our Associates for their support, hoping that they find in EPS a working partner with whom to address many topical issues, and not merely an institution worth some help.

The well-known "red shift" belongs to the past; our finances are now such that we have been able to support some initiatives with seed money. The amount is still very modest, but it hopefully signals a new trend. Electronic publishing benefited in 1992 and east-west activities in 1993 — activities which will be reported upon in Nice. It is also clear that all the important new actions which we initiate have to rely on external funding granted on a case-by-case basis. We are used to this. In general, despite the success, the financial situation cannot be considered as brilliant; it will require much care for some time to come.

EPS in Europe and in the World

I found great stimulation in the challenges which had to be met. The first important challenge was the new orientation which had to be given to east-west activities. Another was to improve relations with the Commission of the European Communities (CEC) and with the European Science Foundation (ESF). Important steps forward could be made. We now collaborate fruitfully with the ESF in organizing the European Research Conferences (ERC) in physics; nine took place in 1992 and ten are prepared for 1993, and the events received a very good review in a recent appraisal by the CEC. Another challenge was to promote relations with The American Physical Society (APS) and with the two physical societies in Japan. There was always a great willingness to foster relations and, in the former case, joint actions on east-west matters were implemented, while permanent contacts were established at the Division and Committee levels. The fourth challenge was the all too familiar financial one which has already been covered.

Wonderful at EPS is to meet colleagues from many different countries, and to discover all that we have in common and all that we can do together. I was fortunately able to visit Belgium, Czechoslovakia, Germany, Greece, France, Hungary, Italy, Poland, Spain, Sweden, and the United Kingdom for professional reasons. Discussions with representatives of national societies could be included, and I strongly felt the welcome extended to me everywhere. My only regret is I could not visit all of our member societies. Three important meetings related to east-west activities were opportunities to meet extensively with colleagues from throughout central and eastern Europe. I also had extensive discussions with fellow physicists from the FSU and there were two visits to the APS, one to Japan, and three to DG-XII in Brussels.

The Society is most clearly seen through the many actions of its Divisions, Interdivisional Groups, and Action Committees and *Europhysics News* has already reported on many of the conferences and meetings organized in 1992. EPS offers the legitimacy and an efficient and convenient technical and legal framework which is often required to develop meetings activities. EPS already does a great deal, but much more could, and should, be done. Specialised conferences and topical meetings in Europe have increasingly to be international to be worthwhile; they are by their nature easily accommodated within the

128

EPS structures. It is felt in some fields that an increasing number of conferences has to be on a worldwide basis. Our excellent relations with the APS, the Association of Asian Pacific Physical Societies, and the Japanese societies should be put to use. For example, our ACAPPI physics in industry committee is already considering the possibility of a joint meeting with the Japanese Society of Applied Physics.

We must not slacken up the development of activities. The end of the cold war has surprisingly been met by a rather difficult economic situation, whereas the many new opportunities should herald important developments. In many places in the world, the present recession fuels extremely nationalistic feelings; people sadly rally around ethnic concepts when we know that in our modern world, economic development calls for more cooperation and collaboration between nations. As physicists, we should counter these trends, showing as much as we can all the benefits of increasing the international collaboration which has already borne so many fruits.

East-West Activities

Turning to topical items, I shall focus on east-west activities, involvement in European research, and education. The special importance of the east-west question was recognized by Council in 1992 with the setting-up of the EWCC (East-West Coordination Committee) to take over from the east-west task force. This committee has been as active as the task force. Special thanks go to Eddy Lingeman, the Secretary of both. An important occasion was the "summit" meeting with the APS in Budapest in May. Priorities and joint lines of actions could be spelt out and quickly implemented; they were also formulated in a joint APS-EPS presidential letter which has been widely circulated. The EWCC and the APS Committee on International Scientific Affairs (CISA) have since held a joint meeting (in Amsterdam last Oct.).

The situation in the former Soviet Union presents a new and formidable challenge and has added much to what we were already busy developing for east and central Europe. The EPS's three main priorities involves promoting journal distribution, modern communication methods, and joint initiatives by research groups by circulating information, providing advice, and channelling useful material.

The 1992 Council endorsed the *Journals for Russia* proposal; action followed immediately, the aim being to ensure that about 30 European physics journals continue to arrive in some 50 institutes in the FSU. The seven publishing companies involved all responded generously to our request for significant price reductions. The proposal has been strongly supported by DG-XII and

129

recommended as a very valuable item in the framework of the scientific help granted by the European Communities to the FSU. Everything is now essentially at the political level in the EC Council of Ministers, and we are presently still waiting for the formal implementation of an Association through which help would be granted. While the journals scheme apparently remains at the highest priority, proposals on other topics are piling up in Brussels. We can but hope for a quick outcome as the journal problem is urgent.*

A detailed **directory** of research institutes in eastern and central Europe is now available, thanks to the EWCC's work. Two others, covering the Baltic States and the other republics of the FSU, will be completed soon. Good contacts have been established with The Physical Society of Japan for this fact-finding exercise and, in general, for collaborative efforts in connection with physics in the FSU.

The EPS **Dagstuhl meeting** in August 1991, which brought together representatives from research organizations in the west and from the physical societies in the east, has been recognized as being very valuable. It helped shape the best ways to proceed and it was at the meeting that the "bottom-up" policy could be defined. It is through special extra funding to research collaborations (at the bottom) that help can be most efficiently channelled. It seems pointless to provide help through existing national structures (at the top) which are in a phase of profound restructuring. Our task is to help bring into contact potential collaborators and make better known all existing possibilities, for there are many information barriers which have still to be lifted. A successor to the Dagstuhl meeting is presently being planned by the EWCC for the Baltic region.

It was heartwarming that several newly created, or recreated, **physical societies** immediately expressed their interest in joining EPS. We were happy to welcome four in 1992 and we shall probably welcome three more in 1993. Most are clearly unable at present to pay their full fee, as are some long-standing members. We have to show much understanding. However, while we wish to welcome physical societies with open arms, we cannot bring into EPS too many new non-paying members. I must again turn to the richer societies in the west, as I did last year, asking them to help with some sponsorship for both newcomers and established members. We benefited in 1992 from a SFR 10 000 gift from the Swiss Physical Society and from smaller amounts for fee support for Estonia from the Finnish and Swedish societies. For 1993, we have so far very generous gifts totalling some 22 kSFR from the UK, German and Swiss

---

*It eventually worked out and the scheme is now operative, thanks to funding by INTAS.

Societies. But we remain somewhat short of the SFR 50 000 of uncollected fee income so the fee question remains a problem.

Involvement in European Research

Physics research increasingly relies on large facilities, which in Europe are more and more internationally based. The EPS should increase its role as a forum for discussion when planning such facilities, since each of them calls for some large constituency of users. As reported in *Europhysics News*, EPS has already taken some steps. It should also increasingly provide advice and help for the implementation of international research networks. Our contacts with the CEC on these matters have been very fruitful, but much remains to be done. For example, we should respond to the invitation of DG-XII to undertake studies on the opportunities and needs for research in Europe. The EPS should not only report on what is happening, as *Europhysics News* now does very well, but should also take initiatives. For we have the technical ability and the legitimacy. This is particularly the case in research fields covered by condensed matter, where international panels and advisory bodies often do not exist, as they do in particle physics, for instance.

Our Associates meeting in Grenoble last November, hosted by the ESRF and ILL, was the occasion for a very fruitful forum on the role of large facilities; it was a very successful meeting.* Much preparatory work on large facilities is now also needed at the world level in order to optimise available resources for research. A joint EPS-APS presidential statement urging more cooperation and collaboration between European and the United States, has been widely circulated. More recent contacts should lead, we hope, to Japan becoming associated to the appeal.

Education

There is much concern about physics education throughout the industrialized world. The American and Japanese physical societies have been particularly active in this area. In Europe, initiatives have developed at the national level, but it is clear that improved international collaboration should be very worthwhile; EPS could play an important role. The structure provided by the Action Committee on Education was recognized as inadequate and in order to better meet the important challenge, the creation of an Interdivisional Group on Education will be proposed to Council in Nice. Two components of

---

*In September '94 the EPS organized a conference on "Large Facilities in Physics", which provided a very good survey of the question. The proceedings are published by World Scientific.

this Group already exist. The first is the **European Mobility Scheme for Physics Students** that is being set up in parallel with the EC's Erasmus and Tempus programmes. The warm response of the 1992 Council to the proposed scheme has been followed up by a series of detailed actions, under the leadership of E. Heer, which are now in the final phase prior to the scheme's launch this coming autumn. Over 120 institutions have already joined the scheme.

The second component is a **Forum on Education** authorised by the Executive Committee in June, which is being set up by G. Marx and G. Tibell. Its aims is to address aspects of pre-university level education for which European collaboration would be beneficial. It is widely recognized that the physics community as a whole has an important responsibility in improving the teaching of physics at the secondary level. With restructuring, we also hope to associate with EPS many **Teachers** who are members of national societies, and we should be able to offer something or direct interest to them from among our many activities. The variety of languages and cultures is certainly a challenge, but it is also a great asset since many different experiments are usually carried out at the same time, and everyone should be able to learn from their various outcomes.

The Amaldi Foundation is creating a **prize** for a physics text book to be granted every other year on a European basis. Through its Forum on Education, EPS will be associated with the award of this prize.* Edoardo Amaldi devoted much time to writing physics text books despite his many research and international activities. His vision, enthusiasm and drive were of utmost importance in the birth of several European research organizations. When arguing many years ago for a European space programme, he once said "Nothing can best provide a link between a Sicilian peasant and a Norwegian fisherman than to know that there are in the sky, complicated objects which circulate around us and which in some way, belong to both of them". The hope is that the great-grandchildren of both, along with and many other young people, will awaken to the beauty and challenge of physics, benefiting from the best talent borrowed from our many European cultures. EPS has a role to play in this, and with the support now solicited from Council, education is set to become a domain in which the Society can greatly extend its activities.

Publications

Publishing is an important source of income for some large physical societies. However, EPS cannot capitalise upon the situation within the context

---

*The prize has now been awarded for the first time. It is an Austrian textbook which won among 50 competitors.

of the present European scene, although *Europhysics Letters* provides a very good example of what can be done in association with national societies. Most of the publishing market remains in the hands of powerful publishing companies who knew better than others how to adapt to a changing world a few decades ago. Even if we cannot enter fully the publishing business, we should care about publications in physics and express our views as a society.

Practically, all leading physics publishers in Europe are now Associate Members, so we work with them on long-range programmes such as electronic publishing. But EPS does not yet have the role which it should eventually take on editorial boards, so efforts must be directed accordingly. We grant an EPS "recognized" status to journals which meet international criteria in agreement with the spirit of our society. All leading journals in Europe, particularly those serving an international community, are invited to apply (the scheme is currently evaluating renewals and new applications after the first five years of operation). There are also many physics journals in central and eastern Europe which cannot survive on economic grounds without some co-ordination and some optimization of the fields covered. First contacts have shown that there is much reluctance to undertake the necessary changes — but we should keep trying, for this is the only way to keep in that part of Europe some physics publications which hard-pressed library budgets will still find necessary to accommodate.

Norbert Kroo, the Vice-President, with whom I have greatly enjoyed collaborating on the Executive Committee for several years, will be proposed as President to Council in Nice. I wish him the greatest possible success in tackling our many challenges, for there is still much to be done.

EPS Restructuring

A sound overall basis

Discussed now are the key points behind the restructuring scheme being prepared for Council in March. While many aspects were dealt with last year in Athens, what is new is a detailed assessment of the financial consequences.

The mandate given by Council in Zurich to the Executive Committee in 1991 led to a restructuring document which was accepted in its general spirit by Council in 1992, in Athens. However, a necessary condition for final approval was that the scheme should be put on a sound financial basis. This is now the case and this is the reason why the Executive Committee is considering presenting a restructuring document to Council in Nice next month. The document is presently being finalised now that we have collected answers to the consultative version sent around before Christmas.

The present plan keeps the dual structure of our Society, which is both *federative* in its being a society of national societies, and *fully international*, through the membership of its Divisions, Interdivisional Groups and Committees, where individual physicists work together at the full European level.

What would be the main outcome of restructuring? Three points are worth mentioning at the present time:

1) Membership on the boards of our Divisions, Groups and committees will now be open to all members of the national societies adopting the "full membership" status (membership is presently restricted to the Society's IOM's).

2) The much wider distribution of *Europhysics News* which will eventually follow will make physicists in Europe more informed and concerned about the activities of EPS. This will foster their sense of participation in a pan-European enterprise.

3) EPS presenting itself as a society of some 60 000 members will be even better recognized as a natural and important partner by the international organizations and institutions which play an increasingly important role in European science.

EPS achieves its stature through the framework and the legitimacy which it provides for actions. The hope is that younger physicists will increasingly feel at home within the Society, and find it the proper place to express their views about the need for the new forums, new facilities, and new structures which are increasingly needed at the international level to pursue research. We have tried whenever possible to help active scientists express themselves, and recent examples in nuclear physics and in supercomputing testify to this. Yet EPS will not substitute for the national societies, which are most efficient at some specific levels, but will develop in harmony with them.

According to the document, representation of the physics community at Council will be set up in a balanced way. The plan is to have a split between a national representation through the national Societies, and a fully international one, through the Divisions and Interdivisional Groups. Members of national societies will benefit, if they so wish, from all the rights and privileges of the present IOM's, short of a special representation at Council.

The IOM category is, however, maintained for people wishing to join the Society without being a member of a national society. It is suggested that the IOM category will also be open to members of a national society who declare themselves as supporting members to mark their special interest and concern for what the Society is doing and standing for. Individual members dedicated to the European ideal are indeed vital for our European society. However, the effort which supporting members invest in EPS will mean much more than their direct financial contributions. The increasing importance given to the Divisions and Interdivisional Groups in defining the Society's course of action should channel more strongly this thriving individual spirit.

Implementation is foreseen to be gradual, with some societies joining fairly soon and others wishing to wait a few years with the present status. Nevertheless, approval should be conditional on the same expression of overwhelming interest as the one which prevailed in Athens.

Published in Europhysics News, Vol. 24, pp. 13–15,
© 1993 Bulletin of the European Physical Society.

THE EPS, PHYSICS AND INDUSTRY
−1991−

Basic research requires large amount of funding. It is curiosity driven as it should be. The problems of the world are such that many people are inclined to think more, according to short term benefits and to challenge large spendings in so-called "pure" research. They would prefer that more support should be allocated to that research which is likely to yield practical applications and in particular, marketable ones. However, paraphrasing MacBeth *Who can look into the seeds of time and say which grain will grow and which will not?* All physicists are convinced that research should proceed on a wide front and that any strong dichotomy between so-called "pure" and "applied" research is likely to be misleading. This is a message to pass on to deciders on any possible occasion.

Yet, we have to realize the cost of basic research and show concern for these problems while trying to avoid that too radical choices are made. This is a question about which physical societies have to show some concern and most of them do. It may therefore, be deemed appropriate to conclude this book with a modest essay on the eternal question of *Physics and Industry*. This short address served as the opening talk for the Conference *Physics for Industry, Industry for Physics*, which was held in Cracow in 1991. For the EPS, this was also the occasion to organize a meeting of its associate members in a central European country.

"This conference is sponsored by the European Physical Society and the Polish Physical Society. Whilst the president of the Polish Physical Society, J. Zakrzewski, is unable to attend, I had the pleasure to meet him yesterday in Warsaw and we could discuss matters of common interest. The Chairman of the Scientific Committee is Dr. J. Goedkoop, also Chairman of ACAPPI, The

EPS Action Committee for Applied Physics and Physics in Industry, which has long been acting as a very important interface between the Society and Industry. This is one of the reasons for my being here.

The relation between Physics and Industry is of extreme importance to us at the EPS and this conference is a magnificent opportunity to find new ways to develop this relation. Difficulties speak for themselves. They have to be overcome. I am therefore, very happy to be here to learn with your help what could be the best possible ways to develop more fruitful contacts between Physics and Industry. This conference, with its three-tier structure of invited talks, poster sessions, and industrial exhibition, should be conducive to interesting ideas.

We live an extraordinary time when barriers long put across our continent have almost suddenly been lifted. This is a time of great opportunities. This is also a time of great challenges since difficulties are quick to appear once the great excitement leaves the floor to practical actions. Yet these difficulties have to be overcome. Ever since its beginning, 23 years ago, the EPS has made the point to ignore as much as it was possible, the division of Europe with which we long lived. This gives it now great opportunities for action.

The EPS, as some of you know, has a rather involved structure. On the one hand, it is both a society of societies, grouping presently 27 different National Physical Societies in Europe, and a society of members. On the other hand, it deals with each individual country through the relevant National Society, but acts in Europe at large, through its Speciality Divisions, Committees and Task-forces. Its role is "to promote the advancement of physics in Europe". This includes fostering the links between Physics and Industry.

I must confess that my whole scientific life, so far, has been that of a research physicist in theoretical high energy particle physics, with a permanent affiliation with CERN since 1972. CERN operates the largest particle physics laboratory in the world. It is a place where external users can pursue research, often not possible elsewhere. Poland has been the first former COMECON country to join CERN. This is a very welcome move. Indeed, the strong and talented polish community of particle physicists has long been associated with research at CERN.

The research which I have been enjoying seems to be very remote from any short-range industrial innovation. Yet, having always work in large laboratories, with affiliations with Brookaven, Saclay, SLAC and Fermilab beside my long and present one with CERN, I have been a close witness of the interplay of basic research and industry in the advance of high energy physics. Whilst the motivation which drives the scientist comes first of all from our inborn human curiosity, and this independently of any possible direct application, it is clear that this research, which is by essence at the limit of present knowledge and present technology, is conducive to many technical innovations. Even if they often fall short of the patent level, advances with instrumentation eventually leads to important developments of practical interest. This is the more so now that "farming out" has become the normal way to work, namely, leaving to Industry many involved development works which it can bring to completion more cost effectively than research laboratories. Industry is presently more and more associated with basic research and therefore, in a better position to capitalize quickly on interesting spin-offs.

Basic research, even if it is pursued independently of possible applications, as it should be, also offers a very valuable training ground for many walks of life. Industry, as it develops its high-tech component, is bound to benefit more and more from an influx of young researchers well-trained in basic research. The training through research has indeed become as important as the training for research. Academia needs for its basic research, more talented young researchers than it can actually keep on a permanent basis. Industry needs more and more researchers with a high level of training. In my own field of particle physics, a young experimentalist has to work at the frontier of knowledge, at the frontier of technology, within the very strict scheduling constraints associated with the use of a large facility and within the framework of an international team. This can but be an excellent training for Industry even if the research themes may seem very remote from potential applications. Some theorists and experimentalists are trained into computer wizards. As H. Casimir said, "It is so important to be early in life confronted with research of greater depth, greater difficulty and greater beauty than one will find later during one's career".

Training through research is very valuable in a field such as particle physics which seems very remote from direct practical applications. It is clearly very valuable in domains which are closer to many potential applications such as Condensed Matter Physics, Atomic and Molecular Physics, Quantum Electronics.... All over Europe, one now sees a trend for a shorter thesis, of typically 2 to 3 years. This falls in line with the general training value also associated with basic research. Indeed, in Europe, over half of the trained physicists are now working in Industry. The corresponding figure for Poland is well below that average and this indicates that there is still much room for the development of close links between research and industry which can be generated that way.

One of the important messages which one has to repeatedly pass on to deciders is to avoid any clear dichotomy between basic research and applied research. The value of research has to be assessed according to its excellence. Whilst some oriented research should be encouraged by public funds, funding agencies should not be more convinced than Industry itself of the particular interest of specific research programmes. Yet, we live in a world where physics is not granted the priority which it once had with funding agencies. Physics research is more and more asked to justify itself according to its role for the benefit of Society. This is not always easy when many pressing needs are felt at the same time. We are all convinced of the cultural and long-range practical value of basic research in physics. Many people have however, still to be convinced! The answer lies to a great extent in a better understanding and collaboration between Physics and Industry.

How Can the EPS Act?

It has been acting already, directly through its ACAPPI and less directly through its many contacts with its Associate Members which group research organizations and companies with an important activity in physics. The ACAPPI has organized one of its meeting in connection with this conference. We took that opportunity to organize also here and now, one of the meetings of our Associate Members.

One of the important and clearly visible role of ACAPPI has been the organization of very successful Europhysics Industrial Workshops. Eight of them have taken place already between 1987 and now. Two are to be held in 1992. There is no lack of ideas for new ones! The EPS can provide contacts, goodwill and expertise through its speciality Divisions. Such conferences, of direct interest for research in Industry should develop in number. The interest of the European Economic Community (EEC) and the European Science Foundation (ESF) for topical conferences in physics, with several recent ones held in various fields of basic research, is likely to extend soon to more applied domains. The EPS has played so far, a strong advisory role in this programme and the conferences in physics have been held as part of the activities of its speciality Divisions. It will do its best to help the extension of such conferences to also meet the present and specific needs of research in Industry. The EPS is strongly developing its links with the American Physical Society and the newly created Association of Asian Pacific Physical Societies. This should help for conferences and workshop which would benefit from a world basis.

The EPS has many specialized committees and task-forces, through which it intends to develop further its actions at this so crucial interface between

Physics and Industry. Its structure and credentials make it in a very good position to speak up and to act, with the help of funding agencies in all domains of physics research, and this spaning the whole of Europe.

It can also act favouring the recruitment of physicists in Industry through its efficient circulation of information. More generally, it can help favouring a greater mobility between Industry and Academia, on a short term basis at the student level, and on a longer term basis at the research scientist level. It can help influence Academic curricula so that they better suit needs expressed by Industry.

It can help in defining qualifications for physicists. In some European countries, "being a physicist" is a recognized profession. In others, it is not yet the case, people being classified as either academics or engineers. There are pressing needs to recognize an international professional qualification for physicists, in particular, in domains connected with Health and Safety. The EPS should help in the now needed Europeanization of physicist qualification.

The EPS can also help in bringing physicists working in Academia as consultants to Industry. Its speciality Divisions offer a very good framework for that.

There is obviously much work to be done developing contacts between Physics and Industry in many ways which are mutually beneficial for education, basic research and industrial research. I look forward to this meeting, and to your help, to see more clearly which are the most fruitful lines for action."